SOLAR FUEL GENERATION

SOLAR FUEL GENERATION

Edited by

YATENDRA S. CHAUDHARY

CRC Press is an imprint of the
Taylor & Francis Group, an **informa** business

CRC Press
Taylor & Francis Group
6000 Broken Sound Parkway NW, Suite 300
Boca Raton, FL 33487-2742

Printed on acid-free paper
Version Date: 20161130

International Standard Book Number-13: 978-1-4987-2551-4 (Hardback)

Visit the Taylor & Francis Web site at
http://www.taylorandfrancis.com

and the CRC Press Web site at
http://www.crcpress.com

Printed and bound in the United States of America by
Edwards Brothers Malloy on sustainably sourced paper

I dedicate this book

To my inspirational parents who guided me to where I am, and for their endless support and encouragement. I am honored to have them as my parents.

And my loving little Angels: Anvika and Anushka, who bring a smile to my face even when I am down.

And my wife!

And I hope they will live in a cleaner world powered by green energy!

Contents

Preface

Driven by the ever-growing demand for energy and the rising atmospheric CO_2 level, there is a thriving interest to tap renewable energy sources. The Sun supplies about 7,000 times more energy than the total energy demand on the Earth. Only a very small fraction of abundantly available solar energy is consumed by biological photosynthetic processes to be stored as biomass (carbohydrates and oils) that can be used as fuels. The development of technologies that can capture solar energy and convert and store it in usable form on a massive scale (greater than what is available with existing technologies) is highly desired for a long-term solution to meet the ever-increasing energy demand. Biological photosynthesis provides a blueprint to translate solar energy into energy-rich molecules (such as H_2 and carbohydrates), commonly known as *solar fuels*. These solar fuels have enormous potential to store a high density of energy in the form of chemical bonds that are transportable. There are two direct pathways to generate solar fuels. One is solar H_2 generation and the other is CO_2 photo-reduction to fuels.

This book provides a comprehensive insight into the current scenario, perspectives, and promising state-of-the-art approaches (photocatalytic and photoelectrochemical H_2 generation, PV-water electrolysis, solar-assisted thermochemical H_2 generation, microbial H_2 generation, and photocatalytic CO_2 reduction) being developed to produce solar fuels. It discusses the fundamental concepts that are crucial to design such solar fuel generating devices.

One of the promising approaches to solar H_2 generation is the splitting of water. One way to split water using solar energy is photoelectrochemical (PEC) water splitting. It relies on the semiconductor photoelectrodes that upon harvesting solar radiation drive the water splitting, which produces H_2 on the cathode and O_2 on the anode. Another way is photocatalytic water splitting, in which the photocatalysts are suspended in an aqueous solution. Upon solar irradiation, charge carriers generated in the photocatalyst facilitate water splitting redox reactions at the catalytic sites on its surface. This approach obviates the need of wiring and has a simple and low-cost photocatalytic reactor design. The third approach is PV-electrolysis, which utilizes the electricity generated through a solar photovoltaic (PV) cell to split water in an electrolyzer. Chapter 2 presents the assessment of these three approaches in terms of economic and technological feasibility. These approaches still face the challenge of finding the right materials that surpass the problems of recombination of photogenerated charge carriers, stability, cost, and, of course, the overall efficiency. Chapters 3 and 4 discuss these challenges and possible strategies to overcome these issues to some extent.

Water splitting can also be driven by high-temperature heat to generate H_2- thermochemical H_2 generation. It consumes only water that generates H_2 and O_2, and the reactants used in this process are reused within each cycle, enabling a closed loop. The high-temperature heat can be provided by solar concentrators. Chapter 5 presents the details of this process.

There are microbes that use solar energy to generate H_2 as a result of their metabolic process. However, the low rate of H_2 production and/or sensitivity to O_2 are causes of concern. Various strategies are being explored for their feasible exploitation to produce H_2. Chapter 6 provides an overview on solar bio-hydrogen generation, including various microbial and enzymatic hydrogen-producing processes, their associated processes and mechanism, and, most importantly, the technological advancements made in this area so far *vis-à-vis* the challenges.

Finding easier and more economical ways to convert CO_2 to energy-rich fuels has been the mainstay of researchers over the past decade. One way to achieve this conversion is photocatalytic reduction of CO_2 to energy-rich fuels such as methanol or methane. It is similar to the photocatalytic solar H_2 generation, except for the fact that CO_2 replaces water as the substrate for photocatalysis. The existing set-up that is already efficient in transportation of natural gas and liquid fuels, in particular, makes those CO_2 reduction products even more attractive. Two major challenges, selectivity and energetics, are preventing this domain of research from developing from its infancy. These vital issues are addressed along with the state-of-the-art process in Chapter 7.

This book will be useful for graduate students and researchers (chemists, physicists, and material scientists) who are engaged in energy research to provide an overview and state-of-the-art approaches to produce solar fuels that appear promising to meet long-term energy requirements of our planet.

This project would not have been possible without the collective contribution of many people. First and foremost, I express my sincere thanks to all the authors. I also thank Prof. B. K. Mishra for his support and encouragement, and my students Biswajit, Kamala, Asim, Aditya, and Smruti for assisting me while editing this book. I also thank the team of Taylor & Francis for their support throughout this project: Iris Fahrer, Alex Edwards, B. Sundaramoorthy, Shikha Garg, and especially Aastha Sharma.

Yatendra S. Chaudhary
CSIR-Institute of Minerals and Materials Technology, Bhubaneswar, India

Editor

Dr. Yatendra S. Chaudhary is a senior scientist at CSIR-Institute of Minerals and Materials Technology, Bhubaneswar, India, and a faculty member at the Academy of Scientific and Innovative Research (AcSIR), New Delhi, India. Dr. Chaudhary earned his PhD for his research on nanostructured photocatalysts for solar-driven water splitting in 2004. Subsequently, he moved to Tata Institute of Fundamental Research (TIFR), Mumbai, India, where he carried out research in materials chemistry. He designed enzyme–semiconductor based photocatalysts for visible light driven CO_2 reduction and H_2 production while working at University of Oxford, United Kingdom. His research accomplishments in the area of nanomaterial and solar fuel research have earned him recognitions such as Green Talent-2011 Award from the Federal Ministry of Education and Research (BMBF), Germany, CSIR-Young Scientist Award-2013 in Chemical Sciences section from the Council of Scientific and Industrial Research, India, and the prestigious Marie Curie Fellowship by the European Union. Dr. Chaudhary has been leading various projects funded by Ministry of New and Renewable Energy, Council of Scientific and Industrial Research, and Science and Engineering Research Board, New Delhi, mainly focusing on the design of artificial photosynthesis devices to produce fuel using solar radiation. He is also a reviewer for many Royal Society of Chemistry and American Chemical Society journals and is a member of the editorial boards of the *Journal of Nanoscience* and *International Journal of Photoenergy*.

His research activities are focused on various facets of colloids and materials chemistry for solar fuel generation. The major focus is on designing the heterostructured photocatalysts with desired morphologies and size to exploit the advantages associated with nanomaterials (such as quantum confinement effects and surface area), hetero-interface with appropriate energetics, and layered structure of semiconductors for efficient solar H_2 generation and CO_2 reduction to fuels.

Contributors

Yatendra S. Chaudhary
Colloids & Materials Chemistry Department
CSIR-Institute of Minerals & Materials Technology
Bhubaneswar, India

and

Chemical Science Division
Academy of Scientific and Innovative Research (AcSIR)
New Delhi, India

Richard Foulkes
Department of Chemical Engineering
University College London, Torrington Place
London, United Kingdom

Chaoran Jiang
Department of Chemical Engineering
University College London, Torrington Place
London, United Kingdom

Naresh Kumar
Department of Geological Sciences
School of Earth, Energy and Environmental Sciences
Stanford University
Stanford, California

Biswajit Mishra
Colloids & Materials Chemistry Department
CSIR-Institute of Minerals & Materials Technology
Bhubaneswar, India

Kamala Kanta Nanda
Colloids & Materials Chemistry Department
CSIR-Institute of Minerals & Materials Technology
Bhubaneswar, India

Sanak Ray
Environment and Sustainability Department
CSIR-Institute of Minerals & Materials Technology
Bhubaneswar, India

Debasis Saran
School of Minerals, Metallurgical & Materials Engineering
Indian Institute of Technology
Bhubaneswar, India

Randhir Singh
School of Minerals, Metallurgical & Materials Engineering
Indian Institute of Technology
Bhubaneswar, India

Pratiksha Srivastava
Environment and Sustainability Department
CSIR-Institute of Minerals & Materials Technology
Bhubaneswar, India

Junwang Tang
Department of Chemical Engineering
University College London, Torrington Place
London, United Kingdom

Balasubramanian Viswanathan
National Centre for Catalysis Research
Indian Institute of Technology—Madras
Chennai, India

Asheesh Kumar Yadav
Environment and Sustainability Department
CSIR-Institute of Minerals & Materials Technology
Bhubaneswar, India

1

Photosynthesis to Solar Fuels: An Introduction

Yatendra S. Chaudhary

CONTENTS

1.1 Introduction

The supply of green and sustainable energy is indispensable and is the most crucial scientific and technological challenge of the twenty-first century. As of today, the major source of the energy comes from fossil fuels such as oil, coal, and natural gas (Figure 1.1). Among these fossil fuels, oil is primarily consumed for energy conversion followed by coal and natural gas.

The demand of the energy is increasing rapidly and is expected to increase twofold by the mid-century as compared to that of current global energy consumption due to increasing population and economic growth. In principle, the energy demand could be met by fossil fuel resources, as these are being formed constantly in Earth's crusts. However, we are using these fossil fuels at a much faster rate than that of the rate they are being formed, and thus fossil fuels are not replenishable at the rate they are being consumed. The fossil fuel on burning generates CO_2, a greenhouse gas. There is a growing concern about the increasing CO_2 emission, which traps heat, consequently leading to steady rise of the Earth's temperature and adversely affecting our health, environment, and climate (Figure 1.2). In 2015, the UN organized Paris Climate Conference, where 196 parties (195 countries and the European Union) reached to an international climate agreement to

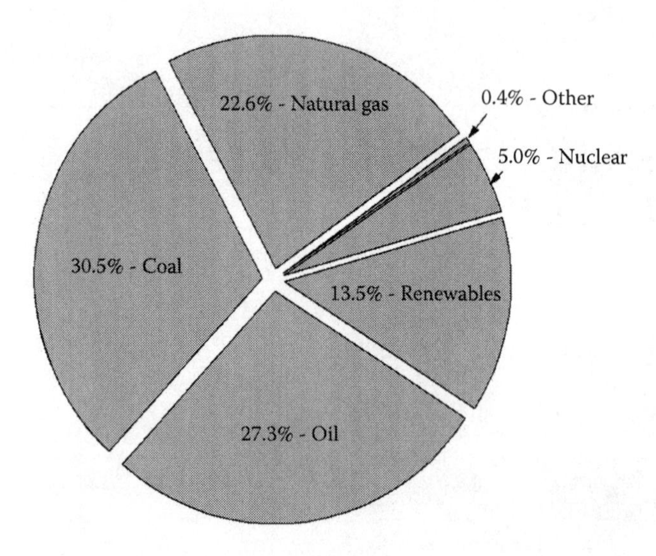

FIGURE 1.1
Energy sources contribution in total global primary energy supply. (Modified from IEA, Renewables Information, 2015. Copyright (2015) International Energy Agency.)

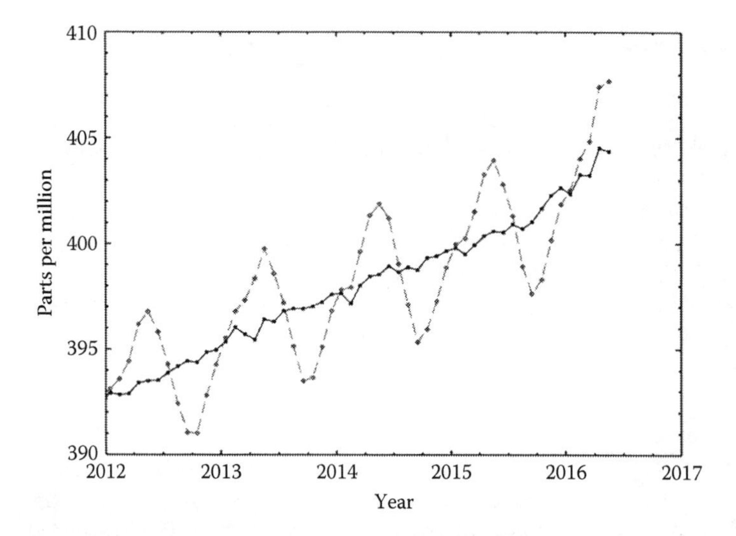

FIGURE 1.2
The monthly mean carbon dioxide measured at Mauna Loa Observatory, Hawaii. The dashed line with diamond symbols shows the monthly mean value and the solid line with a square symbol shows the same after correction for the average seasonal cycle. (Adapted from the U.S. National Oceanic and Atmospheric Administration, 2016. Retrieved from http://www.esrl.noaa.gov/gmd/ccgg/trends/index.html on June 25, 2016.)

limit the temperature rise to below 2°C between now and 2100. Therefore, there is an urgent need to stabilize the CO_2 emission and move away from the fossil fuels as a major energy source. It requires inventions, technological innovation to generate carbon-neutral energy from the renewable energy sources on the scale commensurate with the current energy supply provided by all non-renewable sources.

Most of the renewable energy (wind, tidal, hydroelectric, biomass, etc.) comes directly or indirectly from the Sun, and is intermittent. Among these renewable energy sources, solar energy is by far the largest to meet the increasing global energy demand. There is one more exploitable massive energy source, that is, nuclear fusion, which of course is not renewable, and it is difficult to come up with a technology breakthrough to generate energy at a scale commensurate with the current energy demand. If we look at nature, it has evolved an intricate process, that is, photosynthesis, to produce energy by harvesting solar radiation (solar energy) and store it in the form of chemical bonds (chemical energy). Let us take a look at the intricacies of the biological photosynthetic process.

1.2 Biological Photosynthesis

The biological photosynthesis requires raw materials such as H_2O, solar energy (light), and CO_2 that are available in unlimited amounts. A typical schematic of the photosynthesis process that involves light reactions is shown in Figure 1.3. The photosynthesis consists of three molecular processes: (1) the absorption of light by the antenna that leads to the electron–hole pairs (excitons) generation, (2) the charge separation of excitons and subsequent migration to appropriate reaction sites via various redox-active cofactors located in the photosystems II and I, and (3) redox reactions in which electrons drive the fuel-forming reactions at the reduction catalyst site and holes drive H_2O oxidation at the oxidation catalyst site at which they are quenched by a reductant.

1.2.1 Molecular Processes

Chlorophyll and other pigments absorb solar energy and transfer it efficiently to the photosystem II reaction site, where the charge separation takes place. This conversion of solar energy into electrochemical potential occurs in photosystem II with the thermodynamic efficiency of approximately 70%, generating a radicle pair state $P680^{\bullet+}$ $Pheo^{\bullet-}$ (P680 is chlorophyll a and Pheo is pheophytin a molecule). The $P680^{\bullet+}$ is highly oxidizing with a redox potential of about +1.2 V, whereas $Pheo^{\bullet}$ has sufficiently negative redox potential, that is, −0.5 V. The reducing equivalent is transferred to the photosystem I, where its

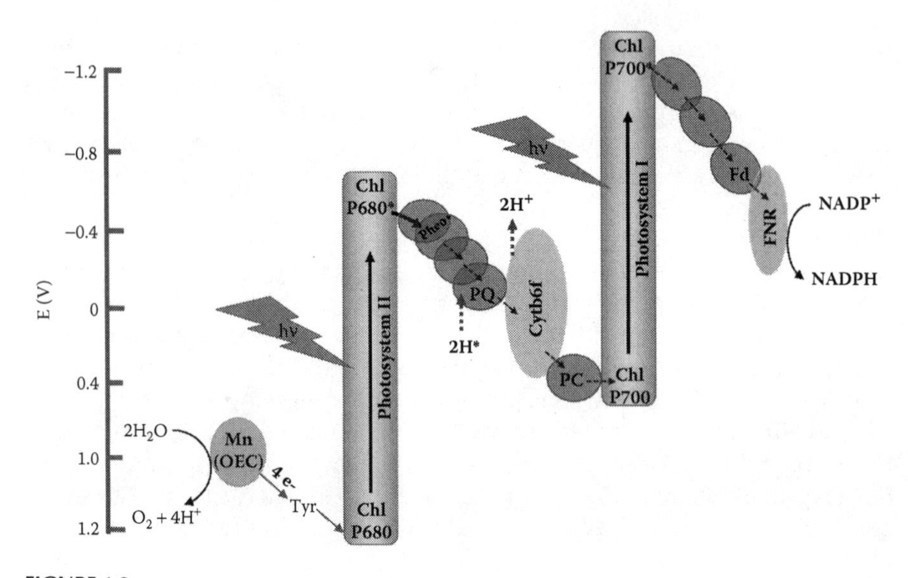

FIGURE 1.3
A simplified depiction of the molecular processes (light reactions) takes place in the biological photosynthesis. (OEC, oxygen evolution catalyst; Tyr, tyrosine; Pheo, pheophytin; Chl, chlorophyll; PQ, plastoquinone; Cyt b_6f, cytochrome b_6f; PC, plastocyanin; Fd, ferredoxin; FNR, ferredoxin-NADP oxidoreductase enzyme.)

reduction potential is raised to –1 V or more, on excitation by the "red" photon absorption at P700 center (chlorophyll molecule). Thus, it gathers sufficient energy to drive the CO_2 fixation reactions. The energy-rich adenosine triphosphate and hydrogen carrier—nicotinamide adenine dinucleotide phosphate—are formed by the energy provided by the chemiosmotic potential generated during electron transfer from photosystem II to photosystem I. Subsequently, they drive the dark reactions at the reactive center, where CO_2 is reduced in carbohydrates—energy-rich organic compounds (fuel). The $Pheo^{\bullet+}$ drives the water splitting reaction while extracting electrons from the Mn-based oxygen evolution catalysts at the oxidation catalyst site. As a result, O_2 is released in the atmosphere (Barber and Andersson 1994; Barber and Tran 2013).

The electrons/protons from H_2O to CO_2 are transferred via various redox-active cofactors present in the photosystems II and I. The transfer of reducing equivalents from photosystem II to photosystem I is assisted by Cyt b_6f. These photosystems and cofactors are arranged in the photosynthetic membrane in a manner that electron flows from water to $NADP^+$ vectorially, leading to the generation of proton gradient along the electron transport chain. In fact, nature has optimized the organization of photosystems and cofactors to facilitate forward fuel-forming and/or energy-storing reactions and curtailing backward reactions (Blankenship 2002).

1.2.2 Efficiency

The photosynthetic organism harvests solar energy throughout the visible solar radiation. However, the energy used to oxidize water and reducing CO_2 is equivalent to that of the red region of solar spectrum. The energy is lost by various processes taking place in the photosynthetic organism that includes the degradation of high-energy photons to the energy equivalent to red photons (i.e., 1.8 eV) by dissipating heat through internal conversion within the solar energy harvesting pigment, overpotentials, driving various reactions such as metabolism, reproduction, and to maintain their organization and survival. While considering all these factors, the maximum efficiency of the photosynthesis is estimated to be about 4.5% (Archer and Barber 2004; Blankenship et al. 2011; Bolton and Hall 1991; Zhu et al. 2010). In general, the efficiency in plants achieved is about 1%–2%, and it is rare to achieve the efficiency of 4%–5%.

1.3 Solar Fuels

There is an enormous amount of solar energy available to us on this planet, in the form of solar radiation. The solar energy reaching to the Earth's surface on an annual basis is ~100,000 TW, which is far greater than that of the current global energy consumption (i.e., 16.3 TW) by the mankind for a year (International Energy Agency 2012). Some of the solar energy is tapped indirectly such as thermal convection (wind farms) and ocean currents (waves and tides). And a very small fraction of abundantly available solar energy (i.e., 0.02%) is consumed by biological photosynthetic processes, to be stored as biomass (carbohydrates, oils) that can be used as fuels. There is a massive amount of solar energy that remains to be tapped. If we take into account the total Earth's surface area, that is, 51,000,000 km^2 and annual solar radiation reaching to the Earth's surface, the global energy demand would be satisfied with an area of 1,150,925 km^2 if the solar energy converted into usable form of energy with 10% efficiency. It amounts to 0.02% of the total surface area or 0.07% of land surface (Larkum 2012). On the other hand, the daunting fact is that the solar radiation is intermittent and diurnal. The development of technologies that can capture solar energy, convert it, and store it into usable form of energy at a massive scale (greater than that of available with existing technologies) is highly desired for a long-term solution. This is where the technologies that convert solar energy into electrical energy do not score for a viable long-term option. The biological photosynthesis provides a blueprint to convert solar energy into energy-rich molecules (such as H_2 and carbohydrates), commonly known as *solar fuels*. These solar fuels have enormous potential to store high density of energy in the form of chemical bonds and are transportable. If the technological breakthroughs are achieved to generate solar fuels with viable efficiency, affordable cost, and stability, they may assist in enabling other sustainable technologies such as carbon capture, storage, and utilization.

Various promising approaches to generate solar fuels are being explored by researchers around the globe. However, they are in their infancy and yet to see the light of day for their exploitation on an industrial scale. There are two direct pathways to generate solar fuels. One is solar hydrogen generation and the other is CO_2 photoreduction to fuels. These are discussed briefly in the following sections.

1.3.1 Solar Hydrogen Generation

Hydrogen has higher energy density (120 MJ/kg) than that of any other fuel and generates environment friendly H_2O on its combustion. These properties make H_2 an ideal energy carrier. Other advantages associated with H_2 are as follows: it can directly be used in existing internal combustion engines/turbines, and it can be used as fuel in fuel cells to produce electricity and also be used for heating applications. Various approaches being explored for solar H_2 generation are as follows.

One way is to split water into H_2 and O_2 using solar energy–photoelectrochemical (PEC) water splitting. In photoelectrochemical water splitting, the solar energy harvested by a semiconductor having appropriate band gap and band edges leads to the generation of charge carriers (an electron is excited to conduction band, leaving behind a hole in the valance band), which subsequently drives the water oxidation on anode and proton reduction to H_2 on cathode. This combines conversion of the solar energy into electrical energy and subsequent electrolysis of water in a single step.

An alternative approach is photocatalytic water splitting, in which the photocatalysts (semiconductor coupled with catalyst) are suspended in aqueous solution. Upon irradiation with solar radiation, charge carriers generated in the semiconductor facilitate water splitting redox reactions at the catalytic sites on its surface. The electron donor (usually water) reduces the photogenerated holes, enabling continuous fuel generation. This approach obviates the need of wiring and has a simple and low-cost photocatalytic reactor design.

In another approach, the electrical energy generated by photovoltaic by harvesting solar radiation is used for water electrolysis to generate H_2. The detailed scientific and technological insight of aforementioned approaches is discussed in Chapter 2. These approaches still face the challenge of finding the right materials that can surpass the problem of recombination of photogenerated charge carriers, stability, cost, and of course overall efficiency. Chapters 3 and 4 discuss about these challenges and possible strategies to overcome these issues to some extent.

The water splitting can also be driven by the heat at a very high temperature to generate H_2. These processes are highly endothermic in nature and are termed as the thermochemical routes to H_2 generation. Cyclic thermochemical processes based on the redox metal oxide, sulfur-iodine, hybrid sulfur cycles, etc. consume only water as a material input and generate H_2 and O_2. The reactants used in this process are regenerated within each cycle, enabling a closed loop. The high-temperature heat can be provided by solar concentrators that can also be used to reform hydrogen-containing fossil/biomass in a noncyclic thermochemical process. Chapter 5 presents the nitty-gritty of such processes.

There are microbes that use solar energy to produce hydrogen during their metabolism. There has been significant progress in exploiting them to produce H_2. However, the low rates of H_2 production and/or sensitivity to O_2 are the causes of concern. Various solar biohydrogen-generating processes range from the light-dependent, direct-biophotolysis, photofermentation, to the light-independent, dark fermentation, microbial electrolysis. Besides, the photobiohybrid devices (enzyme coupled with light-harvesting component usually semiconductors) and the integration of the two types of processes are also gaining interest. An overview on the solar biohydrogen generation, including various microbial and enzymatic H_2-producing processes, their associated processes and mechanism, and most importantly the technological advancements made in this area so far vis-à-vis challenges are discussed in Chapter 6.

It is worth mentioning here that H_2 being the lightest element, its storage is challenging. To realize the potential of the solar hydrogen technologies to the fullest for long-term applications (especially in mobile transportation), further development of commercially viable hydrogen storage technologies is required alongside. It is not discussed further as it is beyond the scope of this book.

1.3.2 CO_2 Photoreduction to Fuels

Finding easier and economical ways to converting CO_2 to energy rich fuels has been the mainstay of researchers over the last decade. One way to achieve this conversion is photocatalytic reduction of CO_2 to energy-rich fuels such as methanol or methane. The existing setup that is already proficient in the transportation of natural gas and liquid fuels, particularly, makes those CO_2 reduction products even more attractive. The methods of photocatalytic CO_2 reduction can be categorized broadly into two classes, photoelectrochemical (PEC) and photocatalysis, which follow almost similar principles as discussed above for solar hydrogen generation. The photogeneration and diffusion of excitons follow exactly the same mechanism in both the cases (solar hydrogen generation and photocatalytic CO_2 reduction to fuels), whereas surface reactions are different in the case of photocatalytic CO_2 reduction. As compared to PEC, photocatalytic CO_2 reduction carries greater prospects due to its simpler functional operation. A variety of photocatalysts are being designed on the similar line of biological photosynthesis blueprint. These photocatalysts can be categorized as (1) semiconducting (heterogeneous), (2) molecular (homogeneous), and (3) heterostructured consisting of two catalysts (one semiconducting and other can be either of a different semiconductor or a molecular). In spite of it, the highest reported rates of product formation are very low and are in the range of only few tens of micromoles per hour of illumination using 1 g of photocatalyst. The two major challenges retarding the progress are selectivity and energetics. The strategies to overcome these challenges and an overview of state-of-the-art photocatalytic CO_2 reduction pathways and their mechanisms are discussed in detail in Chapter 7.

Although the above-discussed approaches are being explored in the laboratory or bench scale, they have promising potential to meet long-term energy requirements of our planet. It can be realized from the following chapters.

References

Archer M. D. and J. Barber. 2004. Photosynthesis and photoconversion. In *Molecular to Global Photosynthesis* (eds., M. D. Archer and J. Barber): 1–41. London: Imperial College Press.

Barber J. and B. Andersson. 1994. Revealing the blueprint of photosynthesis. *Nature 370*: 31–34. doi:10.1038/370031a0.

Barber J. and P. D. Tran. 2013. From natural to artificial photosynthesis. *Journal of the Royal Society Interface 10*, no.81 (April). doi:10.1098/rsif.2012.0984.

Blankenship R. E. 2002. Molecular mechanisms of photosynthesis. Oxford: Blackwell Science (January). doi:10.1002/9780470758472.fmatter.

Blankenship R. E. et al. 2011. Comparing photosynthetic and photovoltaic efficiencies and recognizing the potential for improvement. *Science 332*, no.6031 (May): 805–809. doi:10.1126/science.1200165.

Bolton J. R. and D. O. Hall. 1991. The maximum efficiency of photosynthesis. *Photochemistry and Photobiology 53*, no.4 (April): 545–548. doi:10.1111/j.1751-1097.1991.tb03668.x.

IEA. 2015. *Renewables Information 2015*. IEA, Paris. doi:http://dx.doi.org/10.1787/renew-2015-en.

International Energy Agency. 2012. *Key World Energy Statistics 2012*. Paris, France. http://www.iea.org.

Larkum A. W. D. 2012. Harvesting solar energy through natural or artificial photosynthesis: scientific, political and economic implications. *In Molecular Solar Fuels* (eds., T. J. Wydrzynski and W. Hillier): 1–19. Cambridge: Royal Society of Chemistry.

Trends in Atmospheric Carbon Dioxide. 2016. Earth System Research Laboratory, U.S. NOAA, June 2016. http://www.esrl.noaa.gov/gmd/ccgg/trends/index.html.

Zhu X.-G., S. P. Long, and D. R. Ort. 2010. Improving photosynthetic efficiency for greater yield. *Annual Review of Plant Biology 61* (June): 235–261. doi:10.1146/annurev-arplant-042809-112206.

2

Devices for Solar-Driven Water Splitting to Hydrogen Fuel and Their Technical and Economic Assessments

Richard Foulkes, Chaoran Jiang, and Junwang Tang

CONTENTS

2.1 Introduction

With the growing technology advances as well as the ever-increasing population in this booming world, the global energy consumption rate is expected to increase by a factor of 2, from 15 TW/year today to 27 TW per/year by 2050. According to this trend, it is further expected to increase to 43 TW/year by 2100 (Lewis and Nocera 2006). At present, the main energy supply is obtained from the combustion of fossil fuels, contributing 85% of the total global energy (Barber 2009). However, the upcoming depletion of fossil fuels and linked environmental issues such as pollution and greenhouse gases emission while burning them are the biggest technological challenges being encountered by mankind. Therefore, it is imperative to seek alternative energy supplies to cope with the problem of energy crisis and climate change. Each year, solar energy reaches to the Earth's surface at the annual rate of 100,000 TW of energy, out of which 36,000 TW is on land. This means only 1% of the land is needed to be covered with 10% photoelectrochemical (PEC) cells to generate the energy of 36 TW/year, which is sufficient for the annual energy consumption in 2050 (Barber 2009). Hence, the ability of utilizing solar energy is of great importance for humans. Solar energy can be harvested by using photovoltaic (PV) cells, photocatalysis, and PEC cells to produce hydrogen from water. Hydrogen is environmental-friendly and has three- to four-fold higher mass energy density compared to other fossil fuels (van de Krol 2012). Among these three technologies to split water, PV electrolysis utilizes electricity generated through a coupling with solar PV cells to split water in an electrolyzer. Photocatalytic and PEC water splitting belongs to the direct solar water splitting branch of hydrogen production technologies. Direct solar technologies use incident sunlight to drive the splitting process and are commonly applied using two different system architectures. One system uses photoactive particles suspension on the surface of which reactions occur. The other uses either photocathode or photoanode to produce H_2 and O_2 separately (PEC cells). This chapter is going to compare these three technologies in terms of economy and feasibility, and then detail the most suitable technology for future hydrogen generation by solar-driven water splitting.

2.2 Fundamental Knowledge of Water Splitting

2.2.1 Water Splitting Chemistry

Water splitting reaction is an uphill reaction, which requires minimum Gibbs free energy of 237 kJ/mol. The half-reactions are different depending on whether the electrolyte used is alkaline (Equations 2.1 and 2.2) or acidic (Equations 2.3 and 2.4).

Alkaline electrolyte

$$4H_2O + 4e^- \rightarrow 2H_2 + 4OH^- \quad (2.1)$$

$$4OH^- + 4h^+ \rightarrow 2H_2O + O_2 \quad (2.2)$$

Acidic electrolyte

$$4H^+ + 4e^- \rightarrow 2H_2 \tag{2.3}$$

$$2H_2O + 4h^+ \rightarrow 4H^+ + O_2 \tag{2.4}$$

Therefore, the overall water splitting reaction can be expressed in Equation 2.5, and the potential difference (ΔV) between water oxidation and reduction reaction is 1.23 V, which is also the minimum thermodynamic requirement of solar-driven water splitting.

$$2H_2O \rightarrow 2H_2 + O_2 \Delta G = 237\,kJ/mol \tag{2.5}$$

2.2.2 PV Electrolysis of Water

PV electrolysis combines the electrochemical principles of electrolysis with the photochemical utilization of incident solar energy. In the electrolysis process, a direct current is circulated through water between two electrodes (the anode and the cathode) physically separated by a diaphragm or membrane (Bockris et al. 1981). The electrodes are submerged in water, often with an electrolyte that increases the ionic conductivity. An oxidation reaction occurs at the anode, generating oxygen and causing electrons to flow on to the external circuit—leaving the anode positively charged. The electrons flow to the cathode, negatively polarizing the electrode and producing hydrogen through a reduction reaction. The two half-reactions combine to give the overall water splitting reaction. Separating the electrodes serves to prevent the recombination of hydrogen and oxygen, thereby minimizing the loss of solar energy. A graphical representation of general electrolysis processes is shown in Figure 2.1.

Figure 2.2a shows how the two systems are coupled in an autonomous layout. Being the more mature technologies, hybrid systems involving alkaline (Brinner et al. 1992; Galli 1997; Garcia-Conde and Rosa 1993; Lehman and Chamberlin 1991) and polymer electrolyte membrane (PEM) (Atlam 2009; Clarke et al. 2009; Paul and Andrews 2008; Rashid et al. 2015; Shapiro et al. 2005) electrolyzers have been the focus of much of the research in this area. However, work on solid-oxide electrolyzers has also been carried out providing a proof of concept (Petrakopoulou et al. 2016; Steeb et al. 1985).

Many textbooks provided a detailed explanation of the principles and mathematics behind PV cells (Krauter 2007; Luque and Hegedus 2011; Markvart 2000; Wenham et al. 2009). A general overview of the operation principles will however be briefly covered here (Figure 2.2b). The most common types of PV cells are made using semiconductor materials in crystal lattices such as silicon. The lattices are doped with a species that, relative to silicon, are either electron rich or electron deficient—becoming what is known as n-type and p-type materials, respectively. When these two materials are brought together, they form a p–n junction. The relative difference in electron distribution causes diffusion of electrons from the n-side

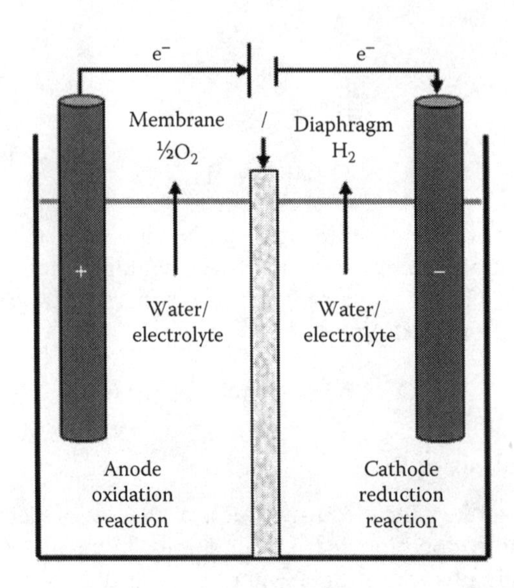

FIGURE 2.1
General representation of electrolysis processes.

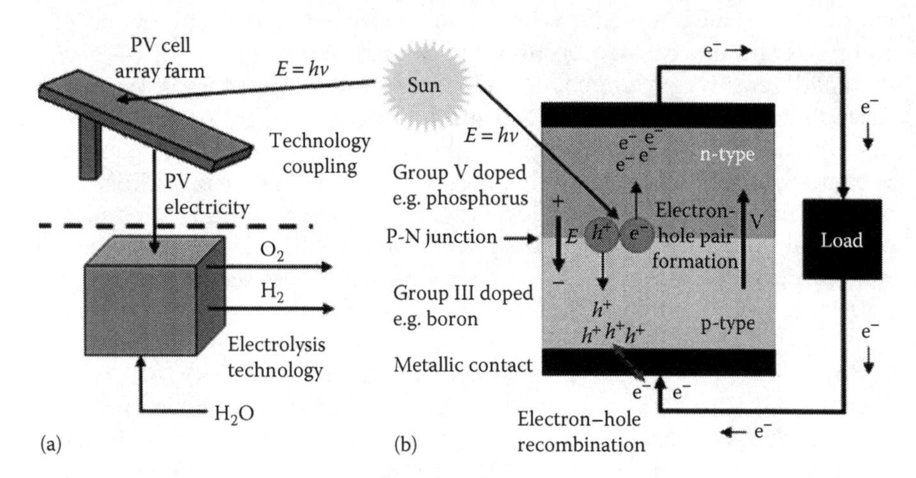

FIGURE 2.2
(a) PV electricity and electrolysis technology coupling. (b) Photovoltaic cell operation.

of the junction across to the p-side. Similarly, holes, attracted to the negative charge, move in the opposite direction from the p-side to the n-side. This charge movement establishes an electric field, E, which builds until it balances the flow. However, the flow is balanced by the electric field acting in the opposite direction. To do this, the field must attract negatively charged electrons to the n-side and positively charged holes to the p-side. Therefore, when incident light rays with sufficient photon energy create electron–hole pairs, liberated electrons move to the n-side and holes move to the p-side.

The separation of these electrons and holes creates a potential, V. When the n-type material is connected via an external load to the p-type material, liberated electrons are attracted by the positive charge and are drawn through. An electric current is thus generated. The electrons reaching the p-type material recombine with the migrating holes, hence restoring charge neutrality. If the external load is the water electrolysis system, electricity generated by the solar cell is utilized to drive water splitting, thus storing solar energy into H_2 fuel.

2.2.3 Photocatalytic and PEC Water Splitting

Technically, a photocatalytic reaction (suspension system) includes three main processes as shown in Figure 2.3. First, it begins with the absorption of a solar photon in a semiconductor material to generate an excited electron–hole pair (Figure 2.4). In order to achieve photocatalytic water splitting using a single semiconductor, the electrons in the conduction band must have more negative potential than 0 V (vs normal hydrogen electrode [NHE] at pH = 0) to conduct the water reduction reaction, and the holes in the valence band must have more positive potential than 1.23 V (vs NHE at pH = 0) to conduct the water oxidation reaction (Figure 2.4). On this basis, the minimum band gap energy of 1.23 eV is required. In practice, larger band energies are required to drive photocatalytic reactions due to energy losses associated with the overpotentials of water reduction and oxidation reactions (Pinaud et al. 2013). The second process is the charge separation and transportation (Figure 2.3). Since photogenerated electron–hole pairs will easily recombine before migrating to surface, in a photocatalytic reaction, sacrificial reagents are commonly employed. For example, a hole scavenger such as triethanolamine (Martin et al. 2014a) or methanol (Jiang et al. 2016) can be used in the water reduction half-reaction to produce hydrogen. Similarly an electron scavenger such

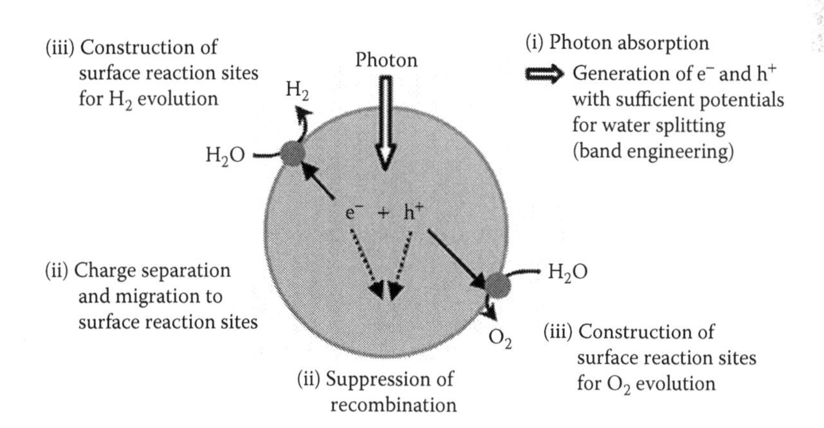

FIGURE 2.3
Three processes in photocatalytic water splitting. (Kudo, A., and Y. Miseki, *Chemical Society Reviews*, 38, 1, 253–278, 2009. Reproduced by permission of The Royal Society of Chemistry.)

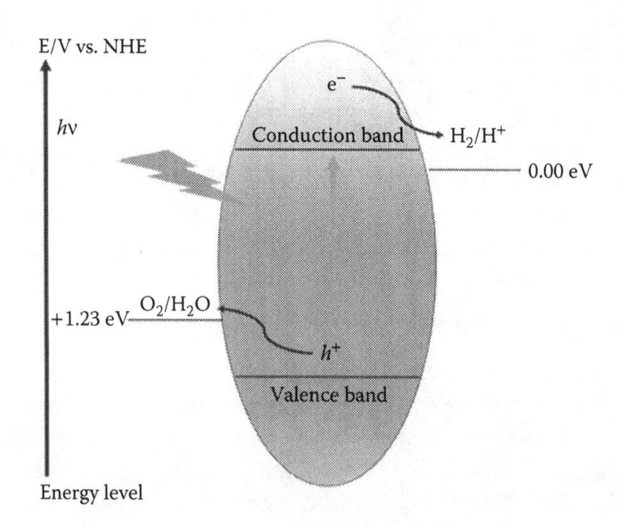

FIGURE 2.4
Photocatalytic water splitting through single semiconductor.

as silver nitrate (Martin et al. 2013) can be used in the water oxidation half-reaction to generate oxygen. The third process is the surface chemical reaction. In order to facilitate the charge separation and surface reaction, photocatalysis is often modified with the appropriate cocatalyst.

The main component of PEC water splitting devices is semiconductor photoelectrode instead of photocatalyst powder. In addition, an electrolyte is necessary. As shown in Figure 2.5, similar to photocatalytic reaction, a complete PEC water splitting reaction contains three processes as well. However, in a PEC configuration, an electrical bias is used to assist electron–hole separation. In addition, O_2 and H_2 can be produced separately on two different electrode surfaces in a PEC water splitting system.

The overall efficiency (solar-to-hydrogen [STH] efficiency) of a PEC water splitting cell is limited by the efficiency of each step. Therefore, the solar-to-hydrogen (η_{STH}) efficiency can be expressed in the following equation:

$$\eta_{STH} = \eta_{abs}\eta_{Sep}\eta_{cat} \tag{2.6}$$

STH efficiency can also be defined as the ratio of total energy generated to total energy input by sunlight illumination:

$$\eta_{STH} = \frac{\text{Total energy generated}}{\text{Total energy input}} = \frac{\Delta G \times R_{H_2}}{P \times A} \tag{2.7}$$

where:

ΔG is the Gibbs free energy (237 KJ/mol)
R_{H_2} is the rate of hydrogen production in mole/s

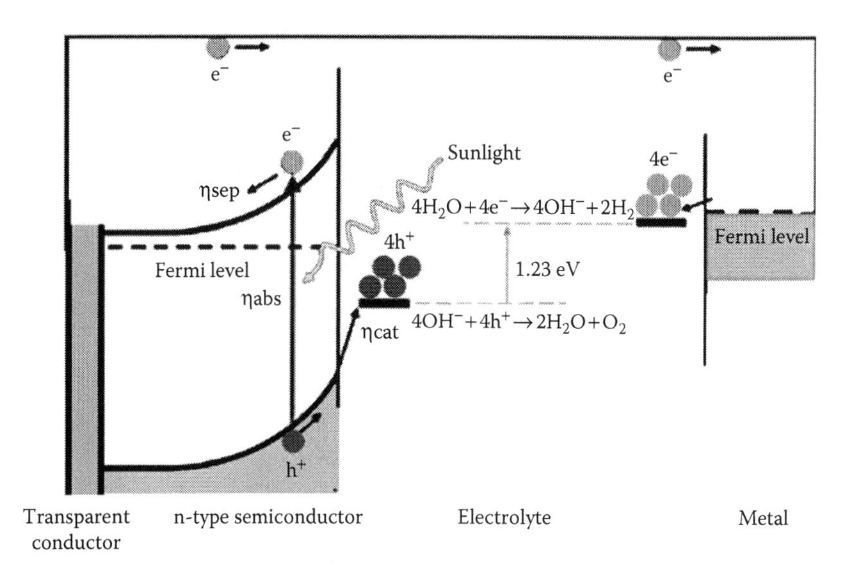

FIGURE 2.5
Schematic diagram of a simple PEC cell and three involved processes.

P is the light intensity (100 mW/cm^2)
A is the illuminated area of the photoelectrode (cm^2)

However, in most cases, the following equation is used to calculate the STH in a PEC cell (Walter et al. 2010):

$$\eta_{STH} = \frac{P_{out} - P_{in}}{P_{light}} = \frac{J_{ph}\left(V_{redox} - V_{bias}\right)}{P_{light}} \times \eta_{Faraday} \qquad (2.8)$$

where:
V_{redox} is the redox potential for water splitting (1.23 V vs NHE)
V_{bias} is the applied bias added between working electrode and reference electrode
P_{light} is the light intensity (100 mW/cm^2)
$\eta_{Faraday}$ is the Faraday efficiency
J_{ph} is the generated photocurrent density

2.3 Technical and Economic Assessments

Although these processes have the common goal of splitting water into hydrogen and oxygen, the economics of each technology is influenced by distinctly different factors. PV electrolysis depends on developments in

PV cells—both price and efficiency. Photocatalytic and PEC water splitting depend on the reactor design and material development. Although this assessment considers a well-to-gate basis, it is worth pointing out that formidable post processing challenges still remain for solar generated fuels such as hydrogen. Notable hurdles include the storage and transportation of hydrogen as well as safe and commercially viable utilization in fuel cells. Work in these areas is ongoing with incremental improvements driving progress.

2.3.1 PV Electrolysis of Water

The economics of PV electrolysis is very similar to that of grid electrolysis. However, because the electricity is generated "in house" with a PV panel, the economics is not as strongly correlated to the grid electricity price. Economic feasibility is a function of the performance of the system (PV cells and electrolysis units together). This means an improvement in the PV technology is a critical factor, and commercial success relies on the development of the technologies performance. As electrolyzers are relatively mature at the current stage, this discussion will focus primarily on the potential advances in PV cells.

2.3.1.1 Technology Drivers

The fundamental difference between electrolysis and PV electrolysis is the source of the electricity feedstock. The former uses grid electricity, while the latter obtains electricity from PV cells. One of the most important reasons for using photovoltaics is the environmentally responsible and sustainable sourcing of electricity feedstock. Recent awareness of environmental issues and a resource-constrained future has certainly pushed energy company sentiment in the direction of PV. However, for many, unless economical, this alone is not enough to incentivize the use of PV technologies. PV must be an economically competitive technology that requires continued performance development.

Incremental improvements and learning curves in the first-generation (silicon wafer) solar cells have gradually reduced the costs of PV technology. Although efficient, the first-generation solar cells are expensive because they require thick wafers and vacuum processes for film fabrication (Wang et al. 2016). Despite this they are expected to remain the dominant PV technology until 2020 when the second-generation systems will become prevalent (Tanaka 2008). The second-generation (thin film) solar cells are fabricated by the deposition of photosensitive materials such as silicon, cadmium telluride (CdTe), and copper-indium-diselenide (CIS) (Tanaka 2008). Thin films, although at the lower end of the PV technology efficiency spectrum, offer a low-cost option that suits large-scale applications where land cost is not significant (Tanaka 2010). The long-term future of PV devices may however be in the third-generation solar cells that seek to combine the advantageous aspects of the first- and second-generation technologies. Promising methods encapsulated in the third generation umbrella term include multijunction cells, intermediate-band cells, hot-carrier cells, and spectrum conversion (Brown and Wu. 2009). Taken from the literature, Table 2.1 gives a technical summary of PV generations and absorption materials (Abermann 2013). What is essential here is that vast amounts of potential and opportunity exist for innovation and development in the field of photovoltaics. As investment in R&D continues, the technology and the case for using PV for water splitting will continue to strengthen. Another advantage of the PV technology is that it supplies DC electricity that is an ideal coupling with the operational need of an electrolyzer (Conibeer and Richards 2007).

2.3.1.2 Plant Operability

The intermittent nature of solar energy introduces complications to powering electrolysis with PV electricity. In the field of photovoltaics, a capacity utilization factor (CUF) is defined as the ratio of the annual energy delivered to the energy that would be delivered annually under ideal conditions to the plant capacity (Equation 2.9):

TABLE 2.1

Comparison of PV Technology Generations and Main Absorber Materials

	First Generation		Second Generation			Third Generation	
Absorber	Sc-Si	mc-Si	a-Si/μ-Si	CdTe	CIGS	DS(S)C	Organic
Maturity of production (%)[a]	86	89	84	77	80	–[d]	–[d]
2010 market share (%)[b]	39	48	4	7	2	~0	~0
Max cell efficiency (%)[c]	25.0	20.4	12.3	16.7	19.6	11.0	10.0
Max module efficiency (%)[c]	21.4	18.2	10.4	12.8	15.7	–[d]	–[d]
Commercial efficiency (%)	14	14	6/–[d]	11	11	–[d]	–[d]
Absorber thickness (μm)	180–250		0.2–0.35/1–2	2–5	2–3	~10	0.03–0.2

Source: *Solar Energy*, 94, Abermann, S., Non-vacuum processed next generation thin film photovoltaics: Towards marketable efficiency and production of CZTS based solar cells, 37–70, with permission from Elsevier.

[a] The maturity of production is obtained by dividing the maximum commercial module efficiency by the maximum laboratory cell efficiency. To the author, this represents well the degree of utilization of the potential of the respective technology in terms of commercial production.

[b] Share was calculated from annual cell/module shipments in MW (Ardani, K., and R. Margolis, *Solar Technologies Market Report*, U.S. Department of Energy, www.nrel.gov/docs/fy12osti/51847.pdf, 2011.)

[c] Confirmed terrestrial cell/module efficiencies measured under the global AM1.5 spectrum (1000 W/m2) at 25°C (IEC 60904-3: 2008, ASTM G-173-03 global) (Green, M. A. et al., *Progress in Photovoltaics: Research and Applications*, 20, 12–20, 2012.)

[d] No confirmed terrestrial module efficiencies available.

$$CUF = \frac{\text{Annual energy delivered (kWh)}}{\text{Plant capacity (kW)} \times 24 \times 364} \tag{2.9}$$

Although clearly affected by day and night, the CUF is also a function of solar irradiance and weather conditions. The CUF of systems where there are generally clear and sunny days, such as the Nevada desert in the USA, has been demonstrated to be around 20% (Liu et al. 2015b). Practically this means that an electrolyzer would only be operational for 1/5th of the year and subject to regular variability in load. While complicated, this is not necessarily prohibitive as one technoeconomic analysis points out (Mason and Zweibel 2008). They found that the cost penalties to hydrogen production for CUF of 25% were only 11% higher than for systems with a capacity factor of 95%. Higher costs in the 25% CUF alkaline electrolyzers considered in this study were offset largely by factors such as reductions in operation and maintenance (O&M) expense and increases in operational life. However, research suggests that PEM electrolyzers would be more appropriate for this type of intermittent operation as they have displayed stable performance during dynamic operation (Millet and Grigoriev 2013).

2.3.1.3 Technoeconomic Findings

Previous research has attempted to simulate PV-electrolysis operations (Balabel et al. 2014; Bezmalinović et al. 2013; Gibson and Kelly 2010; Muller-Langer et al. 2007; Paola et al. 2011). A gate-to-wheel technoecononomic assessment from 2008 however provides the most comprehensive insight into this technology's future (Mason and Zweibel 2008). It argues that because PV module life (approximately 20–30 years) is shorter than balance of plant (BOP) life (approximately 60 years), a single PV plant can be constructed across two generations (note these generations are different to the generations of PV technology that are also referred to). After the PV modules from the first generation have expired, they can be replaced with PV modules that cover the next 30 years of operation. The capital expenditure for the second generation is much reduced because the BOP components remain from the first generation. This study therefore assesses the cost of hydrogen production and distribution across a first- and second-generation facility.

The first-generation plant assesses the cost of hydrogen production and distribution for 10%, 12%, and 14% efficient PV module systems with a 591,780 kg/day capacity. The second generation assumes technological improvement in efficiency over the 30 years of the first-generation plant and therefore assesses 12%, 14%, and 16% efficient PV modules. In addition, the authors posited and investigated that as a relatively new technology thin-film PV cells may have operational lifetimes as long as 60 years, an idea that recent research supports (Abermann 2013). A summary of the results from the 2008 analysis is shown in Table 2.2. The main finding for the first

TABLE 2.2

Results for PV-Electrolysis Hydrogen Production Plant Analysis

	PV Module Efficiency					
	First Generation			Second Generation		
	10%	12%	14%	12%	14%	16%
	Cost of Hydrogen ($/kgH₂)					
20-year PV life	4.89	4.34	3.95	2.78	2.51	2.23
30-year PV life	4.67	4.12	3.75	2.60	2.34	2.12
60-year PV life	–	–	–	1.83	1.68	1.50
	Levalized PV DC Electricity Prices (¢/kWh)					
20-year PV life	7.2	6.1	5.3	4.3	3.8	3.3
30-year PV life	6.4	5.4	4.7	4.0	3.5	3.1
60-year PV life	–	–	–	2.6	2.3	2.1

Source: *Solar Energy*, 94, Abermann, S., Non-vacuum processed next generation thin film photovoltaics: Towards marketable efficiency and production of CZTS based solar cells, 37–70, with permission from Elsevier.

generation plant is that 14% efficient PV modules are required to achieve the Department of Energy's (DOE's) 4–2 $\$/kg_{H2}$ target price. For all efficiencies and across all PV-cell life expectancies, the second-generation plant met the DOE production cost target and even exceeded it for the 60-year case. The cost of electricity was again found to be the largest contribution to hydrogen production costs, accounting for over 80% of the total; thus the hypothesis that focusing on the expenditure of solar cell instead of electrolysis cell is reasonable. Because of this, a sensitivity analysis was conducted for both the overall hydrogen cost and PV electricity generation. The base case was taken as a first-generation plant with a 30-year operating life and 12% efficient PV modules. The result of the analysis (Figure 2.6) showed that the cost of hydrogen was most sensitive to PV-electricity cost by over a factor of 10 relative to other assessed variables. A 1 ¢/kWh change in the cost of PV electricity causes a 0.54 $\$/kg_{H2}$ change in the cost of hydrogen, whereas electrolyzer O&M expense, the next most sensitive variable, causes a 0.053 $\$/kg_{H2}$ change for a 0.5% change in O&M expense. The cost of PV electricity was most sensitive to PV module efficiency and insolation level at 0.4 ¢/kWh for 1% and 0.5 h (21 W/m^2) changes, respectively. These results clearly demonstrate how integral the PV technology is to the commercial success of a PV-electrolysis plant. However, the study assumed that the sites would receive 271 W/m^2 of solar insolation for 6.5 h daily as a base case that is high relative to the global average. Such requirements for high solar insolation levels therefore implicitly exclude many global regions from using this technology.

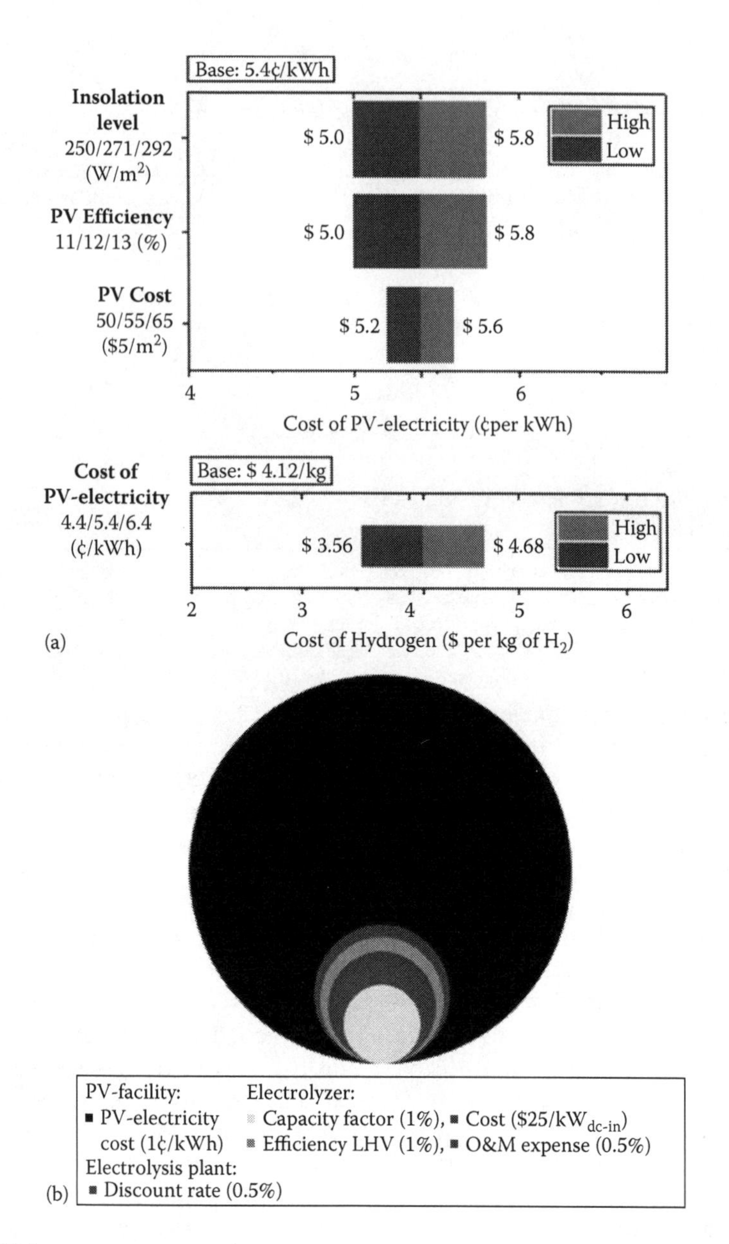

FIGURE 2.6
(a) Tornado charts produced from results for hydrogen and PV-electrolysis production cost sensitivity analysis. (Data from Mason, J., and K. Zweibel, *Solar Hydrogen Generation: Toward a Renewable Energy Future*, vol. 1, 273–313, 2008.) (b) Relative (by area) impact of variables on the cost of hydrogen per unit change in the variable.

However, to provide a contrasting approach to the analysis, the PV and electrolysis systems can also be taken in isolation. Analysis of pure electrolysis from grid electricity also assessed the relationship between electricity price and the impact on the cost of hydrogen (James et al. 2013). The overall PV-electrolysis system can therefore also be assessed using the grid-electrolysis assessment based on the delivery price of PV electricity. In the grid-electricity analysis, an electricity price of 6.50 2007¢/kWh or lower was required to reach the DOE 4–2 $\$/kg_{H2}$ target. The electricity prices in Table 2.2 show that, by this method, the hydrogen cost target would be achieved in all but the most pessimistic case. However, contradictory reports expect PV electricity to be significantly more expensive than the 7.2–2.1 ¢/kWh assumed attainable in the PV-electrolysis analysis. The US Energy Information Agency's current projection for the cost of PV electricity is 9.8–19.3 2013¢/kWh by 2020 depending on the site location (EIA 2015). Even projections out to 2050 under similar insolation levels show mixed support for prices in the 7.2–2.1 ¢/kWh range (Mundada et al. 2016; Parrado et al. 2016; Tidball et al. 2010). Furthermore, Table 2.1 shows that the second-generation commercial CdTe and CIGS PV modules are already close to the expected performance standards outlined in the PV-electrolysis analysis. Similar research has also looked less favorably on the economics, concluding that PV electrolysis could be up to 10 times more expensive than grid-powered electrolysis (Acar and Dincer 2014; Muller-Langer et al. 2007). Therefore, while the assessment of sensitivities provides valuable system insight, the predicted hydrogen costs are not necessarily reliable.

2.3.2 Photocatalytic and PEC Water Splitting

As relatively immature technologies, the basic principles of PEC and photocatalytic water splitting were first demonstrated on a TiO_2 photoanode in 1972 (Fujishima and Honda 1972). Because of this technical immaturity, demonstrations of photocatalytic systems are still confined to bench-scale operations (Baniasadi et al. 2012; Jing et al. 2010; Spasiano et al. 2015; Villa et al. 2013). These operations are essential however for exploring reactor materials and providing a basis for future developments.

2.3.2.1 Technology Drivers

Because of its direct utilization of solar energy, photocatalytic water splitting and other PEC processes at suitably high efficiencies have been touted as the "Holy Grail" of renewable energy sources (Bard and Fox 1995). In addition to being sustainable, PEC processes offer an end-to-end solution. In other words, by taking solar energy as the input and directly utilizing it to produce hydrogen, there are no intermediate steps that may produce carbon emissions, involve capital expenditure, or incur losses.

The delivery pressure of hydrogen is also an important aspect of each system. Delivery pressures of 300 psi (205 bar) achieved by compression with suspension systems however are inherent in high-pressure PEC cell operation. As future hydrogen storage solutions are likely to include compression, this can potentially save on some costs and is useful for automotive applications.

As with PV electrolysis, the separation of oxygen and hydrogen evolution means PEC systems can achieve high product purities. The cogeneration of products in suspension system means a separation stage or novel engineering approaches are required to achieve the required purity. The purity is therefore a function of how rigorous these separations are. The standard (98%) purity has been used for analysis that is still useful for applications in fuel cells (Pinaud et al. 2013).

With a low technology readiness level of 1–2 out of 9, PEC systems are still largely unknown. Therefore, one of the key drivers for this technology is the vast research potential that will undoubtedly lead to improved performance. For example, over the years substantial work has been dedicated to quantifying the realistic efficiencies of PEC devices (Bolton et al. 1985; Rocheleau and Miller 1997; Weber and Dignam 1984, 1986). For tandem photocatalytic systems of interest, the maximum theoretical efficiency is thought to be 40%–41% (Hanna and Nozik 2006; Ross and Hsiao 1977). However, considering practical system losses, the maximum obtainable solar-to-hydrogen (STH) efficiency for PEC systems has been calculated as 22.8% (Seitz et al. 2014). Demonstrated efficiencies have however thus far fallen short and languish in the low single figures. Stable materials with STH efficiencies of 1%–2% have been reported, while higher efficiencies of 5% have only been obtained for materials with hour timescale stability (Fabian et al. 2015; Liao et al. 2013; Liu et al. 2015; Wang et al. 2016). Photocatalysis and PEC cells are therefore attractive technologies because, as opposed to mature technologies, it is both obvious where the improvements will come from and plausible that they be achieved.

2.3.2.2 Plant Operability

Because of the reliance on solar energy, utilization and average daily insolation are important factors in plant operability. To account for this, systems must be oversized and provide the rated capacity. Achieving high insolation also affects the design of the reactors and panels as the effective capture of light becomes an influential consideration.

Two types of conceptual reactor design and panel arrangement (Figure 2.7) have become popular working models since their use in the most comprehensive economic assessment of the technologies to date (Pinaud et al. 2013). However, photoreactor and panel design are an area of ongoing study (Pinaud et al. 2013; Xing et al. 2013). Type 1 reactors are transparent, allowing light to penetrate through, and serve to simply contain

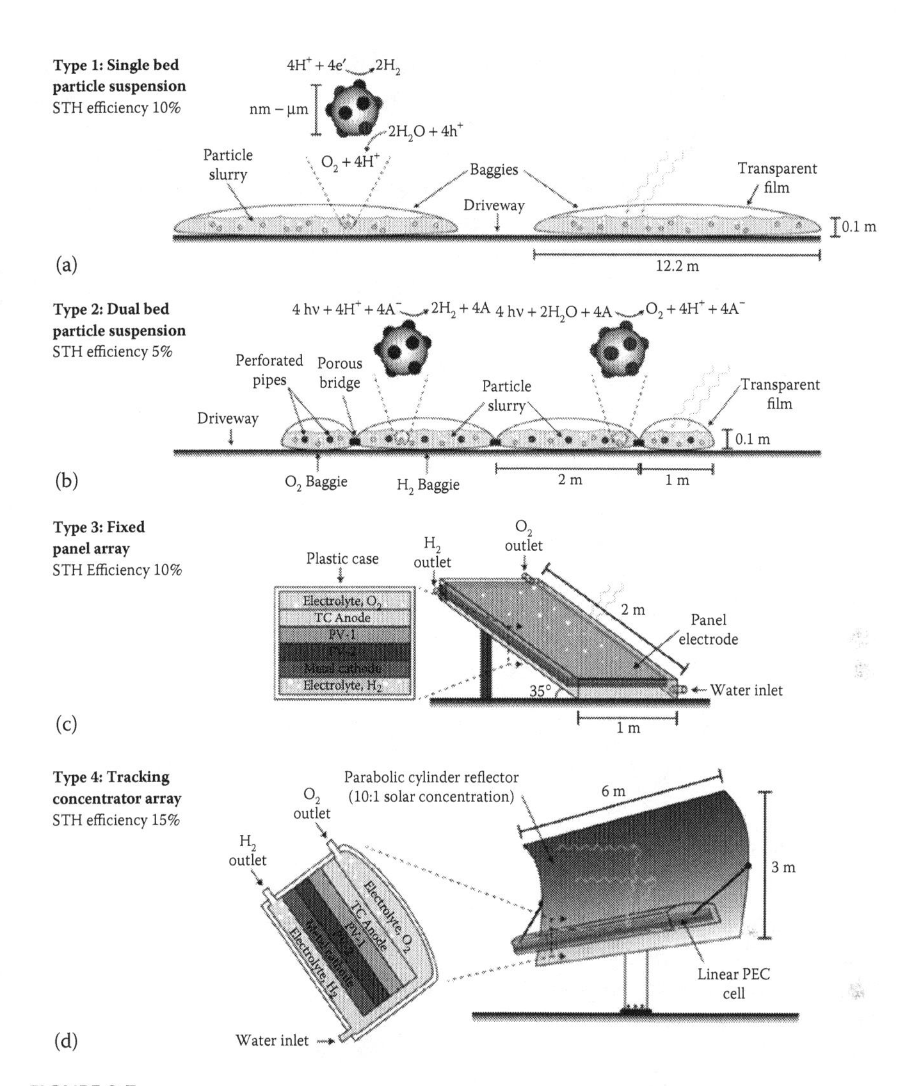

FIGURE 2.7
Schematic of conceptual reactor types with (a) type 1 reactor single baggie cross section with particle slurry, (b) type 2 reactor dual baggie arrangement with separated H_2 and O_2 evolution chambers evolution, (c) type 3 reactor design showing the encased composite panel oriented toward the sun with buoyant separation of gases, and (d) type 4 reactor design with an offset parabolic cylinder receiver concentrating light on a linear PEC cell. (Pinaud, B. A. et al., *Energy & Environmental Science*, 6, 1983–2002, 2013. Reproduced by permission of The Royal Society of Chemistry.)

the evolved gases and photocatalytic particles suspended with an electrolyte as slurry. The cogeneration of gases occurring in type 1 reactors differentiates type 2 reactors that instead evolve H_2 and O_2 in separate beds. As well as removing the need for gas separation units, this affords greater potential to optimize the choice of materials for the oxygen evolution reaction (OER) and the hydrogen evolution reaction (HER). However, this design requires the addition of a redox mediator and porous bridge for transport that introduces additional losses and limitations through media transport rates (Jaini and Fuller 2014). Despite being more complex, the type 2 design avoids the formation and subsequent compression of a combustible hydrogen–oxygen mixture that is a significant safety concern associated with type 1 reactors. Mitigating risks arising from this hazard may yet incur additional control and gas processing costs not captured in this analysis.

In their construction, type 3 reactors resemble the solar panels typically associated with the PV technology. These panels are fixed and so face the equator with a pitch that optimizes performance across the entire year. The assembly includes the two electrodes that sandwich photoactive layers and operate inside a transparent casing, which contains the electrolyte and water. Type 4 reactors have the same structure as type 3 reactors; however, by tracking and concentrating solar radiation, a smaller photocell area is required. Expensive materials with higher efficiency and performance may therefore become economically viable options for these reactor types.

The type 3 and 4 PEC panel reactors are inherently more complex than the type 1 and 2 photocatalytic suspension systems. However, the panel systems are based on a familiar and proven architecture that has been successfully demonstrated with photovoltaics. Although conceptually simple, until prototypes are developed, operability problems will be difficult to foresee with the suspension systems.

2.3.2.3 Technoeconomic Findings

The foremost research into hydrogen production using PEC devices assessed both the technical and economic feasibility of conceptual centralized plants with a 10,000 kg/day capacity (James et al. 2009; Pinaud et al. 2013). Results from this study are summarized in Table 2.3 and shall be discussed in the rest of this section.

At 1.63 $\$/kg_{H2}$, type 1 reactors were found to exceed the DOE's target of 2–4 $\$/kg_{H2}$ and overall provide the most economic arrangement. Type 2 reactors still achieved the target level with a production cost of 3.2 $\$/kg_{H2}$. While this implies robust economic potential, these figures were calculated by assuming baseline STH efficiencies of 10% and 5% for type 1 and type 2 reactors, respectively. Such performance from stable materials has not yet been observed and could be many years from discovery if indeed found at all. However, the purpose of the paper was to establish whether a favorable economic case existed should overcome technical barriers. As this has been confirmed, a stronger incentive exists for pursuing research in overcoming such barriers. The only note one may find is the safety issue in type 1 reactor was not considered here.

The PEC designs were found to have a higher cost of hydrogen production compared to the photocatalytic technology (or suspension system). Type 3

TABLE 2.3

Results of Centralized Hydrogen Production Plant Analysis

	Cost of Hydrogen			
Suspension	**Type 1 Reactor**		**Type 2 Reactor**	
system	**(2005 \$/kg$_{H2}$)**	**% of Total**	**(2005 \$/kg$_{H2}$)**	**% of Total**
Capital costs (direct, indirect, and land)	0.97	59.5	2.27	70.9
Decommissioning costs	0.01	0.6	0.02	0.625
Fixed O&M costs	0.48	29.4	0.8	25
Other Variable costs	0.17	10.5	0.11	3.44
Total	**1.63**	**100**	**3.20**	**100**
	Type 3 Reactor		**Type 4 Reactor**	
PEC	**(2005 \$/kg$_{H2}$)**	**% of Total**	**(2005 \$/kg$_{H2}$)**	**% of Total**
Capital costs (direct, indirect, and land)	8.37	80.8	2.81	69.4
Decommissioning costs	0.07	0.7	0.03	0.7
Fixed O&M costs	1.82	17.5	1.2	29.7
Other variable costs	0.1	1	0.01	0.2
Total	**10.36**	**100**	**4.05**	**100**

Source: James, B. et al., *PEM Electrolysis H2A Production Case Study*, https://www.hydrogen.energy.gov/pdfs/h2a_pem_electrolysis_case_study_documentation.pdf, 2013.

reactors were the most expensive options of all the designs with a cost of 10.36 \$/kg$_{H2}$, while type 4 reactors narrowly missed the 4 \$/kg$_{H2}$ end of the DOE target at 4.05 \$/kg$_{H2}$. Again these findings are based on efficiency and cost performances that have not yet been observed but are deemed obtainable with sufficient research advancement. Being well outside the target range and with no substantial advantage, the extra cost of the type 3 reactor cannot be justified. Designs that concentrate on solar radiation and continually optimize their position through tracking are therefore the likely choice for future designs. Therefore, only type 4 designs will be considered for further analysis of the PEC system.

The results show that capital cost was the most significant contribution to overall costs for all technologies. For the photocatalytic type 1 and type 2 reactors, the total uninstalled hydrogen production costs were 59.5% and 70.9%, respectively (Figure 2.8). For the type 1 reactors, high compression costs are caused due to the extra volume of O_2 processed in the product stream. The intrinsic separation of type 2 reactors means its compression costs are cheaper, although this is more than offset by higher costs in almost every other category. The costs attributed to baggies are in particular higher for the type 2 reactors because so many are required that incurs additional fabrication, land, and labor costs. The simpler type 1 reactor coupled with a pressure swing absorption unit therefore appears more economical.

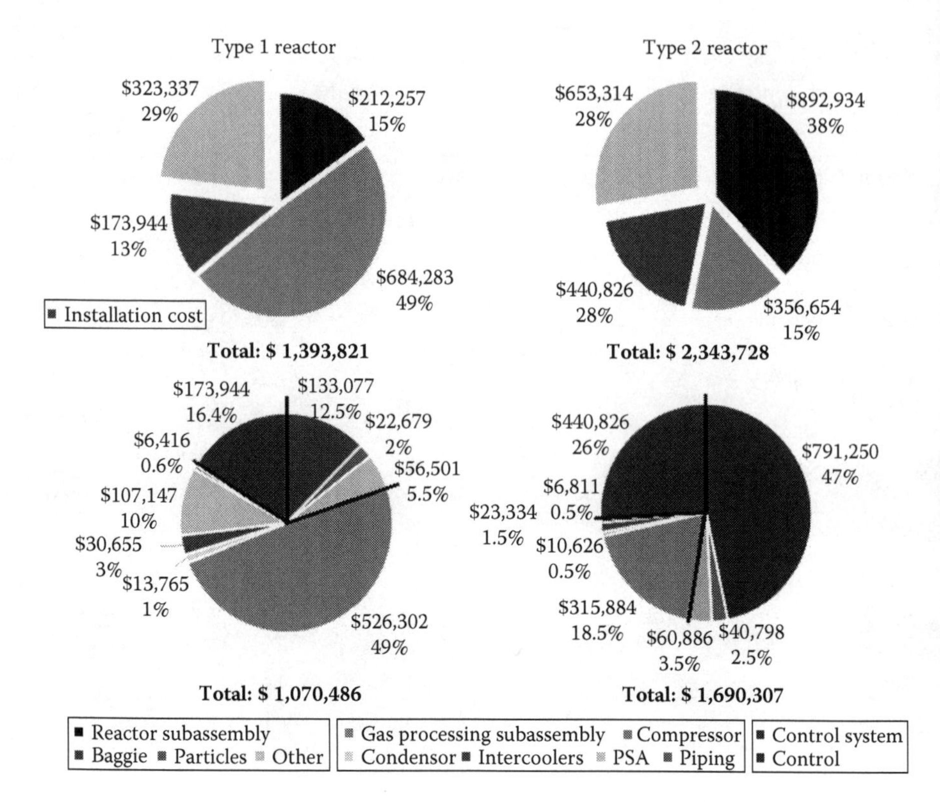

FIGURE 2.8
Breakdown of capital costs (uninstalled and installed) for type 1 and type 2 reactors. (Data from Pinaud, B. A. et al., *Energy & Environmental Science*, 6, 1983–2002, 2013; James, B. D. et al., *Technoeconomic Analysis of Photoelectrochemical (PEC) Hydrogen Production*, 2009.)

Capital costs associated with the panels of the type 4 PEC system accounted for 80.8% of the uninstalled total with fixed O&M the other significant contribution at 17.5%. A further breakdown of the costs (Figure 2.9) shows that the tracking and concentration system as well as PEC cells form the majority of these costs. These can be considered key areas of focus for future research to reducing the costs of these systems. Despite a higher nominal system efficiency of 15%, the uninstalled capital cost and installed total cost of the PEC system is still much higher than the photocatalytic suspension systems. This is almost entirely due to the reactor subassembly that is more costly due to the complex PEC cell design.

To further understand the impact of capital costs and other parameters, a sensitivity analysis was conducted. For the photocalaytic system, the STH efficiency, particle capital costs, and system lifetime were all varied over a feasible range with cell cost replacing particle cost for the PEC analysis.

The Tornado charts for the photocalaytic analysis are shown in Figure 2.10. For both reactor types, the largest effect on hydrogen price was caused by

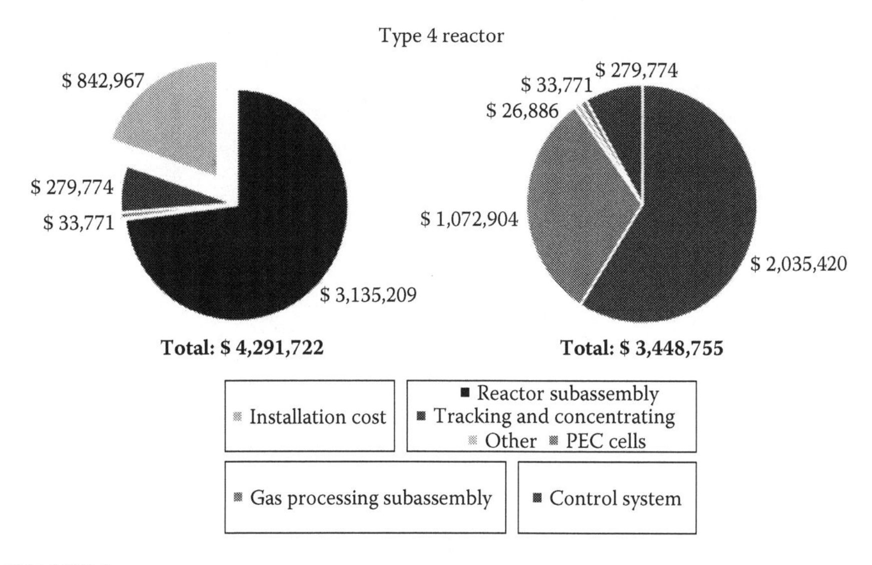

FIGURE 2.9
Breakdown of capital costs (uninstalled and installed) for type 4 reactors. (Data from Pinaud, B. A. et al., *Energy & Environmental Science*, 6, 1983–2002, 2013; James, B. D. et al., *Technoeconomic Analysis of Photoelectrochemical (PEC) Hydrogen Production*, 2009.)

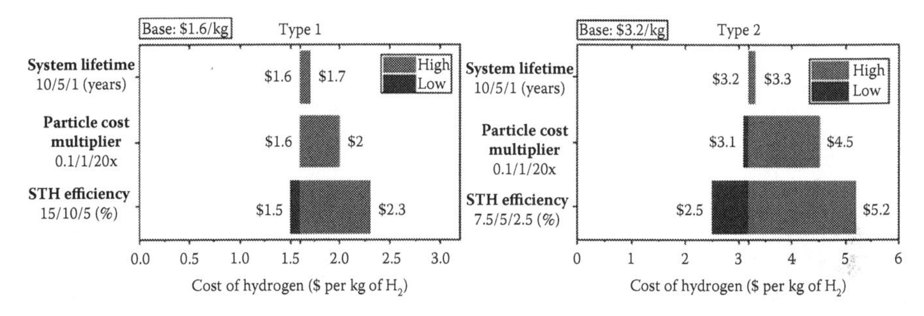

FIGURE 2.10
Reproduced Tornado charts for type 1 and type 2 reactors showing sensitivity to STH efficiency, particle cost, and lifetime. (Pinaud, B. A. et al., *Energy & Environmental Science*, 6, 1983–2002, 2013. Reproduced by permission of The Royal Society of Chemistry.)

changes to the STH efficiency followed by the particle cost multiplier. While hydrogen from type 1 reactors remained within the 2–4 $\$/kg_{H2}$ cost target for both high scenarios, the type 2 reactor was not found to achieve the target for STH efficiencies of 2.5% and particle cost multipliers of 20. As extreme estimates, however, these findings are far from prohibitive and still result in hydrogen costs lower than many other competing technologies. The effect of the system lifetime was found to have a very limited effect for both the high and low cases. The only low case that caused a significant reduction in

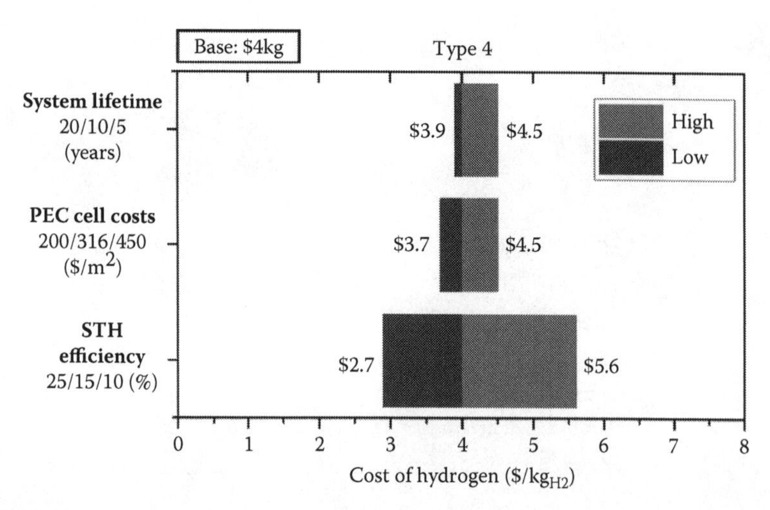

FIGURE 2.11
Reproduced Tornado charts for type 4 reactors showing sensitivity to STH efficiency, PEC cell cost, and system lifetime. (Pinaud, B. A. et al., *Energy & Environmental Science*, 6, 1983–2002, 2013. Reproduced by permission of The Royal Society of Chemistry.)

hydrogen cost was an STH efficiency of 7.5% in type 2 reactors where the cost fell by 0.7 $\$/kg_{H2}$.

The results of a similar analysis for the PEC system are shown in Figure 2.11. Again the STH efficiency has the greatest influence on the overall cost of hydrogen production with a range of 2.7–5.6 $\$/kg_{H2}$ for the 25% and 10% STH efficiency, respectively. Reducing the hydrogen cost by 1.3 $\$/kg_{H2}$, a 25% STH efficiency demonstrated the greatest benefit of any variable across all the technologies. This highlights the importance of the variable to the overall credibility of the system. The sensitivity to system lifetime and cell cost is relatively contained with a 0.1 and 0.3 $\$/kg_{H2}$ benefit respectively in the low case and a 0.5 $\$/kg_{H2}$ penalty for both in the high case. Although the sensitivities technically bring the cost above the target price, the analysis still demonstrates that PEC cells are an economically viable technology. This credibility warrants further research where STH efficiency is a clear priority.

Unlike PV electrolysis, this study did not investigate the effect of solar insolation for either technology. An average insolation level of 219 W/m^2 was assumed that, close to the global average, opens up more potential regions. Future work analyzing the impact of solar insolation on this technology feasibility would be welcome.

Overall the economic case for centralized hydrogen production via photocatalysis is the more robust of the two technologies. The photocatalytic systems appear able to consistently produce hydrogen under the target price for even the high cases. However, safely processing the combustible mixture in type 1 reactors and suitable redox mediators and bridge materials in type 2 reactors remain challenges for the photocatalytic systems. The realization of the particle-bag architecture also remains unknown that may be significant for the future of the photocatalytic systems. The analysis also provides an economic endorsement for PEC systems which, with sufficient development, will also price competitively. Caveats to both systems are that the envisaged technical performance targets, such as STH efficiency, must be reached. However, many of the performance parameters used in the analysis still require substantial research efforts for these systems to be realized. As such the next section will investigate some of the PEC materials with the potential to achieve the required activity and stability. If successful, type 4 reactor will be applied.

2.4 Material Development for PEC Water Splitting

2.4.1 Photoanode Materials

There are many candidates for photoanode materials, and so this section will primarily focus on summarizing the low cost and efficient ones as well as strategies to improve their performance.

1. *$BiVO_4$(BVO)* is an n-type semiconductor consisting of relatively inexpensive and abundant elements. It has a direct band gap of 2.4 eV with a conduction band near 0 V versus reversible hydrogen electrode (RHE) and a valence band at ca. 2.4 eV versus RHE. Therefore, photogenerated holes from $BiVO_4$ have sufficient overpotential to oxidize water. However, an external bias is required to conduct water reduction reactions (Shaner et al. 2015; Sun et al. 2015a, b, c). Theoretically, the maximum photocurrent and STH efficiency of BVO are 7.4 mA/cm^2 and 9.1%, respectively (Li et al. 2013; Wu et al. 2014). The efficiency of BVO is limited by (1) fast charge carrier recombination due to short electron diffusion length of BVO (only 10 nm) (Del Alamo 2011; Zhong et al. 2011) and (2) poor surface water oxidation kinetics. Electron diffusion length can be significantly increased up to ~300 nm by doping BVO with Mo and W with a diffusion coefficient of 1.5×10^7 cm^2/s (Li et al. 2013; Murphy et al. 2006; Radi et al. 2010; Sun et al. 2015), and poor surface kinetics can be modified by surface oxygen evolution cocatalysts such as Co-Pi (Pilli et al. 2011; Zhong et al. 2011). To date, Zhong et al. (2011) have reported a near-zero recombination loss by using W-doped BVO photoanode with Co-Pi modification. Recently, for a single BVO photoanode, a benchmark photocurrent of 2.73 mA/cm^2 at 0.6 V (vs RHE) was achieved on the nanoporous BVO photoanode with a FeOOH/NiOOH dual layer as the OER catalyst (Kim and Choi 2014). By assembling a tandem cell

with silicon, the benchmark efficiency of 4.6% was reached using a BVO-based photoanode (Abdi et al. 2013).

2. *CdS.* Theoretically, overall water splitting under visible light irradiation can be achieved by using CdS due to the small band gap (2.4 eV) as well as a conduction and valence band that straddle the redox potential for pure water splitting (Rabinovich and Hodes 2013). Although it has a long charge carrier diffusion length (up to the micrometer scale), slow water oxidation kinetics leads to accumulation of photogenerated holes at the surface, and thus anodic photocorrosion occurs. Therefore, a stabilization strategy must be applied on similar group II–VI semiconductor materials (e.g., CdS, CdTe, CdSe, ZnTe) for solar-driven PEC water splitting processes. One strategy is to use sacrificial hole scavengers such as EDTA (Darwent and Porter 1981), $Fe(CN)_6^{4-}$ (Reber and Rusek 1986), S^{2-} (Kalyanasundaram et al. 1981), and SO_3^{2-} (Kalyanasundaram et al. 1981) that produce hydrogen. However, in order for the water oxidation reaction to occur on the CdS photoanode, hole scavengers cannot be added to the system. Therefore, another strategy is used whereby an insulation layer is added to prevent from photocorrosion. Recently, Lewis et al. demonstrated that n-type CdTe photoanodes could be stabilized when a 140-nm-thick amorphous TiO_2 by atomiclayer deposition was used together with a thin overlayer of an Ni-oxide-based oxygen-evolution electrocatalyst (Lichterman et al. 2014). Borse et al. reported CdS photoanodes modified with TiO_2 nanoparticles via thioglycerol as an organic linker for long-term hydrogen production (Pareek et al. 2014b), and that CdS photoanodes could also be modified by nanoniobia for an efficient and stable PEC cell (Pareek et al. 2014a). The n-type cadmium telluride photoanode could also be protected by sputtering transparent catalytic nickel oxide (NiO) on the surface that serves not only as a protection layer but also as an efficient oxygen evolution cocatalyst (Sun et al. 2015). With respect to the efficiency, a recent study has shown that a low hole transfer rate was an efficiency-limiting factor in a CdS-based heterostructure (Wu et al. 2014).

3. *III–V compounds.* The III–V compound semiconductors such as GaAs and InP, and quaternary alloys have great potential to photoelectrodes in PEC cells. In general, III–V semiconductors possess several characteristics that make them advantageous for PEC water splitting. (1) Their band gaps are narrow (1.42 eV for GaAs and 1.35 eV for InP) that is ideal for light absorption and as such near-optimal absorptivity of the solar spectrum can be achieved (Sakai et al. 1993). (2) They exhibit extraordinary charge carrier mobility, for example, GaAs has an electron mobility of up to 9200 $cm^2\ V^{-1}\ S^{-1}$, and hole mobility up to 400 $cm^2\ V^{-1}\ S^{-1}$ (Del Alamo 2011). Turner et al. fabricated a monolithic PV/PEC device for hydrogen production via

water splitting with an impressive STH efficiency of 12.6% by using a tandem cell consisting of a p/n GaAs bottom cell connected to a $GaInP_2$ top cell through a tunnel diode interconnect (Khaselev 1998). Although when compared to metal oxides such as TiO_2, Fe_2O_3, and TiO_2, high photocurrent and STH efficiency can be achieved by III–V (Figure 2.12), and the instability and high cost limit the semiconductor applications (Hu et al. 2014). Recently, Lewis demonstrated that GaAs and GaP could be stabilized against photoanodic corrosion or dissolution by the use of a conformal atomic layer deposition (ALD)-deposited TiO_2 layer in conjunction with nickel oxide/hydroxide as an electrocatalyst (Hu et al. 2014; Lichterman et al. 2014; Shaner et al. 2015). The group also reported that GaAs, GaP, and InP could be protected by a multifunctional layer of NiO_x (Li et al. 2015; Sun et al. 2015). The NiO_x acted not only as a protection layer but also as an efficient oxygen evolution cocatalyst.

2.4.2 Photocathode Materials

P-type materials such as cuprous oxide (Cu_2O) (Li et al. 2015), Si (Sun et al. 2014), and SiC (van Dorp et al. 2009) with a conduction band more negative than the redox potential of (H^+/H_2) are considered good candidates for a photocathode. However, due to cathodic photocorrosion in electrolyte,

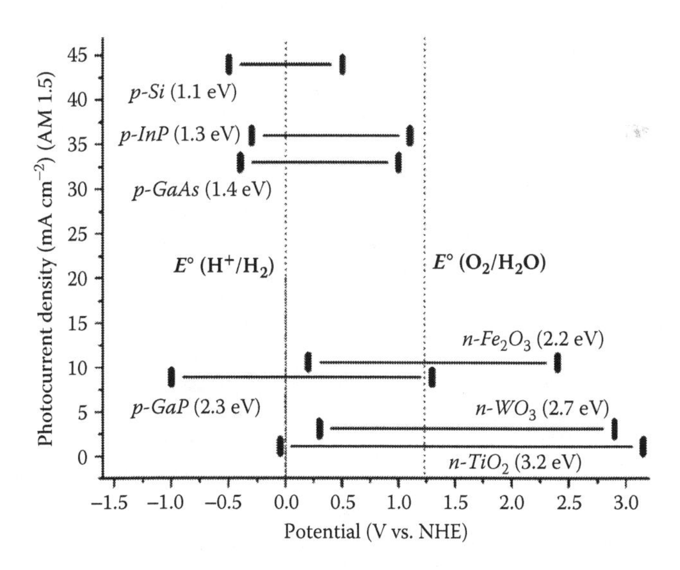

FIGURE 2.12
The band gap position of various semiconductors and their theoretical photocurrent. (Reprinted with permission from Walter, M. G. et al., *Chemical Reviews*, 110, 6446–6473, 2010. Copyright [2010] American Chemical Society.)

their application as a photocathode is still limited (Moniz et al. 2015; Paracchino et al. 2012).

Cuprous oxide (Cu_2O) is a p-type semiconductor with a direct band gap of ~2 eV and suitable conduction band position that enables visible light absorption and hydrogen evolution (Mahmoud et al. 2011; Radi et al. 2010; Wu et al. 2010). A theoretical photocurrent of ~15 mA cm^{-2} and 18% STH efficiency can be achieved by a Cu_2O photocathode under AM 1.5 light based on its band gap (Paracchino et al. 2011). Furthermore, due to the low cost of Cu_2O, large-scale fabrication of the photoelectrode is also attractive. However, the major drawbacks of Cu_2O in solar hydrogen production are the low practical STH efficiency due to the fast electron–hole recombination and poor stability because the redox potentials for the reduction and oxidation of monovalent copper oxide lie within the band gap (Paracchino et al. 2012; Wu et al. 2010). To address these issues, several strategies have been applied on Cu_2O photocathodes including (1) combining with other n-type semiconductors with more positive conduction bands, such as TiO_2 (Wang et al. 2013), ZnO (Jiang et al. 2013), WO_3 (Lin et al. 2012), RGO (Tran et al. 2012), etc. and (2) forming an n–p junction to allow fast transfer of photoinduced electrons from Cu_2O to the conduction band of an n-type semiconductor (Hou et al. 2009). This improves both the efficiency and stability. In addition, the deposition of a thin protection layer on Cu_2O also works. The protection layer could be a thin layer of carbon or NiO_x. Recently, Grätzel et al. reported a $Cu_2O/ZnO/Al_2O_3/TiO_2/Pt$ electrode, resulting in a benchmark photocurrent of –7.6 mA cm^{-2} at 0 V versus RHE with improved stability due to the protective nature of TiO_2 and high conductivity of ZnO/Al_2O_3 (AZO) (Paracchino et al. 2011). The group also designed a similar device consisting of a Cu_2O/n-AZO/TiO_2/$MoS_{2+}x$ heterojunction photocathode that exhibited improved photocurrent and stability in harsh acidic environments (–5.7 mA cm^{-2} at 0 V vs RHE at pH 1.0) (Morales-Guio et al. 2014). Very recently, a Cu_2O-based photocathode of −6.3 mA cm^{-2} at 0 V versus RHE in 1 M KOH electrolyte was achieved by coupling surface protected Cu_2O with a MoS_2 HER catalyst (Morales-Guio et al. 2015). This appears to be the first report of MoS_2 as a highly active hydrogen evolution catalyst in basic medium. MoS_2 thus demonstrates significant potential as an earth-abundant HER catalyst alternative to platinum for water splitting in alkaline conditions.

Silicon has a small band gap of 1.1 eV, which is nearly ideal for use in a dual band gap p/n –PEC water splitting configurations (tandem PEC water splitting devices). A few reviews have summarized that p-Si photocathode combined with cocatalysts can be used to produce H_2 efficiently (Sun et al. 2014). STH efficiency as high as 6% under monochromatic 633 nm illumination has been achieved by a p-Si photocathode embedded with Pt nanoparticles. It is well known that silicon is not stable in aqueous solution under illumination. Therefore, several groups have deposited a corrosion-resistant protective layer on top of p-type and n-type silicon that acts as a photocathode and photoanode for H_2 and O_2 production, respectively.

Ji et al. demonstrated a silicon-based photocathode for water reduction with an epitaxial $SrTO_3$ protection layer and a mesh-like Ti/Pt nanostructured cocatalyst. This resulted in a long-term (35 h) performance in 0.5 M H_2SO_4 with an applied bias photon-to-current efficiency of 4.9%. Very recently, the stabilization of silicon photoanodes in alkaline condition has attracted more attention. Si photoanodes have been stabilized by several novel strategies including (1) deposition of catalytic transition-metal coatings (e.g., CoO_x [Pijpers et al. 2011] and NiO_x [Sun et al. 2015]), (2) deposition of ultrathin metal films (e.g., Ni) on an oxidized Si surface (Kenney et al. 2013), and (3) ALD deposition of a thin amorphous TiO_2 layer between Si and a surface oxygen cocatalyst such as an IrO_2 layer (Chen et al. 2011) or islands of Ni oxide electrocatalyst (Hu et al. 2014; McDowell et al. 2015). Tandem PEC water splitting devices based on multijunction of silicon (p/n Si-PEC) have large potential as they are able to achieve solar-driven water splitting without a bias. In addition, n-type semiconductors such as $BiVO_4$ (Abdi et al. 2013; Shell et al. 2014), WO_3 (Shell et al. 2014), Fe_2O_3 (Wang et al. 2013), and ZnO (Shi et al. 2011) can be coupled with the surface to protect the silicon cell and achieve efficient photocatalytic water oxidation.

2.4.3 Co-Catalyst Selection

Noble metal-based catalysts such as IrO_2 and RuO_2 are among the best oxygen evolution catalysts (OECs), but prohibitively expensive (Ma et al. 2012; Riha et al. 2013). In recent years, abundant, first-row, transition-metal oxides have been demonstrated to be excellent OECs (Yu et al. 2015). For PEC water oxidation cocatalysts, most of these metal oxides are CoO_x or NiO_x (Galán-Mascarós 2015).

For OECs coupled with a photoanode in a PEC cell, cobalt oxide prepared by electrodeposition has emerged as an efficient water oxidation catalyst, which is able to operate under neural conditions (Liao et al. 2014). Nocera et al. prepared cobalt oxide films from phosphate buffer (Co-Pi) (Surendranath et al. 2012) or borate buffer (CoB_i) (Reece et al. 2011) solution at potential above 1.1 V versus NHE (Kanan and Nocera 2008; Galán-Mascarós 2015). The obtained form was amorphous and contained substantial amounts of anions from the phosphate or borate buffer solution. The buffer solution was essential for the deposition of highly active films as well as maintaining their stability. However, the films were unstable in the electrolyte in the absence of an applied bias. The mechanism of these films for water oxidation has been investigated by in situ X-ray absorption spectroscopy, which demonstrated that O_2 evolution was accompanied by an increase in the intensity of a Co (IV) electron paramagnetic resonance signal (Kuang et al. 2016).

Nickel (Ni) is another earth-abundant element from the first transition-metal row. Nickel-based OECs have been widely used in electrochemical and PEC water splitting. Dai et al. demonstrated that a thin layer of Ni metal could be used as an OEC and passivation layer for silicon photoanodes (Kenney et al. 2013). Nickel oxide (NiO_x) prepared by electrodeposition

(Blakemore et al. 2015), sputtering (Sun et al. 2015), and sol-gel process (Shi et al. 2014) has been intensely studied in the literature. However, most of the prepared NiO_x films were converted to either $Ni(OH)_2$ (Wei et al. 2015) or NiOOH during the water oxidation reaction in an alkaline electrolyte (KOH solution). Yan et al. found that the highly nanostructured α-$Ni(OH)_2$ nanocrystals could be a remarkably active and stable OER catalyst in alkaline media. They achieved a current density of 10 mA/cm^2 at a small overpotential of 0.331 V and a small Tafel slope of 42 mV/decade. It was also found that the stability performance of α-$Ni(OH)_2$ nanocrystals was much better than precious RuO_2 (Gao et al. 2014). Naldoni et al. demonstrated that a thin layer of photodeposited amorphous NiOOH coupled onto a hematite (α-Fe_2O_3) photoanode could reduce the overpotential (onset potential shift by 150 mV) and increase the photocurrent by about 50% at 1.23 V versus RHE (Martin et al. 2014). Nocera et al. also reported nickel borate (Ni-B_i), prepared either by electrodeposition or photoelectrodeposition in the presence of nickel precursor (nickel nitrate or nickel chloride) and borate buffer (Dincă et al. 2010), has been widely used to enhance the efficiency of Fe_2O_3 (Klahr et al. 2012), ZnO (Jiang et al. 2014; Steinmiller and Choi 2009), WO_3 (Seabold and Choi 2011), and $BiVO_4$ (Choi et al. 2013). On the other hand, catalysts such as Ni-B_i could undergo self-healing in proper electrolyte (potassium borate solution), which enabled water oxidation over a prolonged period (Bediako et al. 2013; Farrow et al. 2013).

FeOOH has been coupled with various semiconductor materials to act as a photoanode for efficient water oxidation. Mullins et al. deposited α-FeOOH onto Si triple junction solar cells with a photovoltaic efficiency of 6.8%. The obtained photoanode (α-FeOOH/Si) achieved an STH efficiency of 4.3% at 0 V versus RHE in a three-electrode system with 4 h lifetime (Martin et al. 2014b). Ye et al. reported that FeOOH could be loaded onto hematite photoanodes by photoelectrodeposition. The photocurrent of obtained FeOOH/Fe_2O_3 films increased nearly four-fold at 0.43 V (vs RHE), and the onset potential exhibited a cathodic shift by 140 mV compared to a bare hematite photoanode. In addition, long-term stability (70 h) was achieved by Fe_2O_3/FeOOH photoanodes (Ye et al. 2015).

Incorporating Fe into the nickel hydroxide induced the formation of NiFe layered double hydroxide (NiFe-LDH) (Gong et al. 2013; Wang et al. 2015). The NiFe-LDH OEC exhibited higher electrocatalytic activity for oxygen evolution than either NiOOH or FeOOH catalyst (Gong et al. 2013). Ni-Fe LDH could be prepared by a hydrothermal or photoelectrodeposition method onto a semiconductor material. Schmuki et al. introduced an NiFe-LDH cocatalyst onto a Ta_3N_5 electrode by the hydrothermal method. The obtained NiFe-LDH/Ta_3N_5 electrode exhibited photocurrent of 6.3 mA/cm^2 at 1.23 V_{RHE} in 1 M KOH and improved the stability over the test period compared to bare Ta_3N_5 (Wang et al. 2015). Domen demonstrated that nanoworm $BiVO_4$ with Ni-Fe-LDH modification, prepared by photoelectrodeposition, exhibited high STH efficiency (2.25%) and long-term durability

(10 h) in a 1 M potassium borate electrolyte (pH = 9.3) under AM 1.5 one sun illumination (Kuang et al. 2016).

For water reduction, noble metals such as platinum (Pt) are frequently used due to their low overpotential and strong corrosion resistance (Memming 2008). However, efficient inexpensive hydrogen evolution catalysts including metal sulfides, metal selenides, metal carbides, metal nitrides, metal phosphides, and heteroatom-doped nanocarbons are essential for economical H_2 production (Zou and Zhang 2015).

2.5 Conclusion

In summary, the technoeconomic assessment of three methods of producing hydrogen through splitting water has been carried out. For PV-electrolysis plants, the cost of electricity feedstock was found to have the most influence on the overall cost of producing hydrogen. Improving the efficiency of electrolyzers can somewhat reduce this burden, but developing PV technology plays a key role to provide a cheap electricity feedstock. Photocatalysis and PEC cells are still technically immature but indicate promise for the future. Current challenges involve discovering new materials that can prevent electron–hole recombination and utilize a higher proportion of the solar spectrum to increase STH efficiency that dominates the overall costs in the two technologies. Once this is achieved, it is envisaged that capital costs will become a significant, but not prohibitive, expense to reduce.

Hydrogen production from photocatalysis provided the most positive economic case with both type 1 and type 2 reactors producing hydrogen at 1.6 and 3.20 $\$/kg_{H2}$, respectively. Only the type 4 reactor that utilizes PEC technology was deemed sufficiently competitive, producing hydrogen at 4.05 $\$/kg_{H2}$. Although more expensive, the inherent separation of gases and greater certainty of architecture means PEC systems provide a more well-rounded and holistic solution. In addition, sufficient advancements in the STH efficiency of PEC devices can potentially reduce the cost to 2.7 $\$/kg_{H2}$. This would bring the economic performance within the US DOE target and in line with the photocatalytic technology. For these reasons, although both remain interesting prospects for the future, PEC systems are favored. The further analysis on the different components in a PEC cell underlines that the group III–V compounds have a strong potential to meet the target of high STH efficiency, together with an appropriate cocatalyst that also works as a protection layer from photocorrosion. However, other photoelectrodes (including metal oxides, sulfides, and so on) are not ruled out at this stage. In total, a highly efficient PEC device with an affordable cost and stability is the most important in the solar-driven water splitting field.

References

Abdi, F. F., L. Han, A. H. Smets, M. Zeman, B. Dam, and R. van de Krol. 2013. Efficient solar water splitting by enhanced charge separation in a bismuth vanadate-silicon tandem photoelectrode. *Nature Communications* 4 (July): 2195. doi:10.1038/ncomms3195.

Abermann, S. 2013. Non-vacuum processed next generation thin film photovoltaics: Towards marketable efficiency and production of CZTS based solar cells. *Solar Energy* 94 (August): 37–70. doi:10.1016/j.solener.2013.04.017.

Acar, C., and I. Dincer. 2014. Comparative assessment of hydrogen production methods from renewable and non-renewable sources. *International Journal of Hydrogen Energy* 39, no. 1 (January): 1–12. doi:10.1016/j.ijhydene.2013.10.060.

Ardani, K., and R. Margolis. 2011. *Solar Technologies Market Report*. U.S. Department of Energy. www.nrel.gov/docs/fy12osti/51847.pdf.

Atlam, O. 2009. An experimental and modelling study of a photovoltaic/proton-exchange membrane electrolyser system. *International Journal of Hydrogen Energy* 34, no. 16 (August): 6589–6595. doi:10.1016/j.ijhydene.2009.05.147.

Balabel, A., M. S. Zaky, and I. Sakr. 2014. Optimum operating conditions for alkaline water electrolysis coupled with solar PV energy system. *Arabian Journal for Science and Engineering* 39, no. 5 (April): 4211–4220. doi:10.1007/s13369-014-1050-6.

Baniasadi, E., I. Dincer, and G. F. Naterer. 2012. Exergy and environmental impact assessment of solar photoreactors for catalytic hydrogen production. *Chemical Engineering Journal* 213 (December): 330–337. doi:10.1016/j.cej.2012.10.018.

Barber, J. 2009. Photosynthetic energy conversion: natural and artificial. *Chemical Society Reviews* 38, no. 1 (November): 185–196. doi:10.1039/B802262N.

Bard, A. J., and M. A. Fox. 1995. Artificial photosynthesis: Solar splitting of water to hydrogen and oxygen. *Accounts of Chemical Research* 28, no. 3 (March): 141–145. doi:10.1021/ar00051a007.

Bediako, D. K., Y. Surendranath, and D. G. Nocera. 2013. Mechanistic studies of the oxygen evolution reaction mediated by a nickel–borate thin film electrocatalyst. *Journal of the American Chemical Society* 135, no. 9 (January): 3662–3674. doi:10.1021/ja3126432.

Bezmalinović, D., F. Barbir, and I. Tolj. 2013. Techno-economic analysis of PEM fuel cells role in photovoltaic-based systems for the remote base stations. *International Journal of Hydrogen Energy* 38, no. 1 (January): 417–425. doi:10.1016/j.ijhydene.2012.09.123.

Blakemore, J. D., R. H. Crabtree, and G. W. Brudvig. 2015. Molecular catalysts for water oxidation. *Chemical Reviews* 115, no. 23 (July): 12974–13005. doi:10.1021/acs.chemrev.5b00122.

Bockris, J. O. M., B. E. Conway, and E. Yeager. 1981. *Comprehensive Treatise of Electrochemistry*. New York: Springer US.

Bolton, J. R., S. J. Strickler, and J. S. Connolly. 1985. Limiting and realizable efficiencies of solar photolysis of water. *Nature* 316 (August): 495–500. doi:10.1038/316495a0.

Brinner, A., H. Bussmann, W. Hug, and W. Seeger. 1992. Test results of the hysolar 10 kW PV-electrolysis facility. *International Journal of Hydrogen Energy* 17, no. 3 (March): 187–197. doi:10.1016/0360-3199(92)90126-H.

Brown, G. F., and J. Wu. 2009. Third generation photovoltaics. *Laser & Photonics Reviews* 3, no. 4 (July): 394–405. doi:10.1002/lpor.200810039.

Chen, Y. W., J. D. Prange, S. Dühnen, et al. 2011. Atomic layer-deposited tunnel oxide stabilizes silicon photoanodes for water oxidation. *Nature Materials* 10 (June): 539–544. doi:10.1038/nmat3047.

Choi, S. K., W. Choi, and H. Park. 2013. Solar water oxidation using nickel-borate coupled $BiVO_4$ photoelectrodes. *Physical Chemistry Chemical Physics* 15, no. 17 (May): 6499–6507. doi:10.1039/C3CP00073G.

Clarke, R. E., S. Giddey, F. T. Ciacchi, S. P. S. Badwal, B. Paul, and J. Andrews. 2009. Direct coupling of an electrolyser to a solar PV system for generating hydrogen. *International Journal of Hydrogen Energy* 34, no. 6 (March): 2531–2542. doi:10.1016/j.ijhydene.2009.01.053.

Conibeer, G., and B. Richards. 2007. A comparison of PV/electrolyser and photoelectrolytic technologies for use in solar to hydrogen energy storage systems. *International Journal of Hydrogen Energy* 32, no. 14 (September): 2703–2711. doi:10.1016/j.ijhydene.2006.09.012.

Darwent, J. R., and G. Porter. 1981. Photochemical hydrogen production using cadmium sulphide suspensions in aerated water. *Journal of the Chemical Society, Chemical Communications* 4: 145–146. doi:10.1039/C39810000145.

Del Alamo, J. A. 2011. Nanometre-scale electronics with III–V compound semiconductors. *Nature* 479 (November): 317–323. doi:10.1038/nature10677.

Dincă, M., Y. Surendranath, and D. G. Nocera. 2010. Nickel-borate oxygen-evolving catalyst that functions under benign conditions. *Proceedings of the National Academy of Sciences of the United States of America* 107, no. 23 (April): 10337–10341. doi:10.1073/pnas.1001859107.

EIA. 2015. Levelized cost and levelized avoided cost of new generation resources in the annual energy outlook 2015, https://www.eia.gov/forecasts/aeo/electricity_generation.cfm, Accessed March, 2016.

Fabian, D. M., S. Hu, N. Singh, et al. 2015. Particle suspension reactors and materials for solar-driven water splitting. *Energy & Environmental Science* 8, no. 10 (October): 2825–2850. doi:10.1039/C5EE01434D.

Farrow, C. L., D. K. Bediako, Y. Surendranath, D. G. Nocera, and S. J. Billinge. 2013. Intermediate-range structure of self-assembled cobalt-based oxygen-evolving catalyst. *Journal of the American Chemical Society* 135, no. 17 (April): 6403–6406. doi:10.1021/ja401276f.

Fujishima, A., and K. Honda. 1972. Electrochemical photolysis of water at a semiconductor electrode. *Nature* 238 (July): 37–38. doi:10.1038/238037a0.

Galán-Mascarós, J. R. 2015. Water oxidation at electrodes modified with earth-abundant transition-metal catalysts. *ChemElectroChem* 2 (January): 37–50. doi:10.1002/celc.201402268.

Galli, S. 1997. Development of a solar-hydrogen cycle in Italy. *International Journal of Hydrogen Energy* 22, no. 5 (May): 453–458. doi:10.1016/S0360-3199(96)00105-X

Gao, M., W. Sheng, Z. Zhuang, et al. 2014. Efficient water oxidation using nanostructured α-nickel-hydroxide as an electrocatalyst. *Journal of the American Chemical Society* 136, no. 19 (April): 7077–7084. doi:10.1021/ja502128j.

Garcia-Conde, A., and F. Rosa. 1993. Solar hydrogen production: A spanish experience. *International Journal of Hydrogen Energy* 18, no. 12 (December): 995–1000 doi:10.1016/0360-3199(93)90081-K.

Gibson, T. L., and N. A. Kelly. 2010. Predicting efficiency of solar powered hydrogen generation using photovoltaic-electrolysis devices. *International Journal of Hydrogen Energy* 35, no. 3 (February): 900–911. doi:10.1016/j.ijhydene.2009.11.074.

Gong, M., Y. Li, H. Wang, et al. 2013. An advanced Ni–Fe layered double hydroxide electrocatalyst for water oxidation. *Journal of the American Chemical Society* 135, no. 23 (May): 8452–8455. doi:10.1021/ja4027715.

Green, M. A., K. Emery, Y. Hishikawa, W. Warta, and E. D. Dunlop. 2012. Solar cell efficiency tables (version 39). *Progress in Photovoltaics: Research and Applications* 20, no. 1 (January): 12–20. doi:10.1002/pip.2163.

Hanna, M. C., and A. J. Nozik. 2006. Solar conversion efficiency of photovoltaic and photoelectrolysis cells with carrier multiplication absorbers. *Journal of Applied Physics* 100, no. 7 (October): 74510. doi:10.1063/1.2356795.

Hou, Y., X. Y. Li, Q. D. Zhao, X. Quan, and G. H. Chen. 2009. Fabrication of Cu_2O/TiO_2 nanotube heterojunction arrays and investigation of its photoelectrochemical behavior. *Applied Physics Letters* 95, no. 9 (August): 093108. doi:10.1063/1.3224181.

Hu, S., M. R. Shaner, J. A. Beardslee, M. Lichterman, B. S. Brunschwig, and N. S. Lewis. 2014. Amorphous TiO_2 coatings stabilize Si, GaAs, and GaP photoanodes for efficient water oxidation. *Science* 344, no. 6187: 1005–1009. doi:10.1126/science.1251428.

Jaini, R. R., and T. F. Fuller. 2014. Overcoming mass-transfer limitations in the dual bed colloidal suspension reactor. *International Journal of Hydrogen Energy* 39, no. 6 (February): 2462–2471. doi:10.1016/j.ijhydene.2013.12.018.

James, B. D., G. N. Baum, J. Perez, and K. N. Baum. 2009. *Technoeconomic Analysis of Photoelectrochemical (PEC) Hydrogen Production.* https://www1.eere.energy.gov/hydrogenandfuelcells/pdfs/pec_technoeconomic_analysis.pdf.

James, B., W. Colella, M. Jennie, G. Saur, and T. Ramsden. 2013. *PEM Electrolysis H2A Production Case Study.* https://www.hydrogen.energy.gov/pdfs/h2a_pem_electrolysis_case_study_documentation.pdf.

Jiang, C., K. Y. Lee, and C. M. Parlett, et al. 2016. Size-controlled TiO_2 nanoparticles on porous hosts for enhanced photocatalytic hydrogen production. *Applied Catalysis A: General* 521: 133–139. doi:10.1016/j.apcata.2015.12.004.

Jiang, C., S. J. A. Moniz, M. Khraisheh, and J. Tang. 2014. Earth-abundant oxygen evolution catalysts coupled onto ZnO nanowire arrays for efficient photoelectrochemical water cleavage. *Chemistry—A European Journal* 20, no. 40 (September): 12954–12961. doi:10.1002/chem.201403067.

Jiang, T., T. Xie, L. Chen, Z. Fu, and D. Wang. 2013. Carrier concentration-dependent electron transfer in Cu_2O/ZnO nanorod arrays and their photocatalytic performance. *Nanoscale* 5, no. 7 (April): 2938–2944. doi:10.1039/C3NR34219K.

Jing, D., L. Guo, L. Zhao, et al. 2010. Efficient solar hydrogen production by photocatalytic water splitting: From fundamental study to pilot demonstration. *International Journal of Hydrogen Energy* 35, no. 13 (July): 7087–7097. doi:10.1016/j.ijhydene.2010.01.030.

Kalyanasundaram, K., E. Borgarello, D. Duonghong, and M. Grätzel. 1981. Cleavage of water by visible-light irradiation of colloidal CdS solutions; Inhibition of photocorrosion by RuO_2. *Angewandte Chemie International Edition in English* 20, no. 11 (November): 987–988. doi:10.1002/anie.198109871&.

Kanan M. W., and D. G. Nocera. 2008. In situ formation of an oxygen-evolving catalyst in neutral water containing phosphate and Co^{2+}. *Science* 321, no. 5892 (August): 1072–1075. doi:10.1126/science.1162018.

Kenney, M. J., M. Gong, Y. Li, et al. 2013. High-performance silicon photoanodes passivated with ultrathin nickel films for water oxidation. *Science* 342, no. 6160 (November): 836–840. doi:10.1126/science.1241327.

Khaselev, O. 1998. A monolithic photovoltaic-photoelectrochemical device for hydrogen production via water splitting. *Science* 280, no. 5362 (April): 425–427. doi: 10.1126/science.280.5362.425.

Kim, T. W., and K.-S. Choi. 2014. Nanoporous $BiVO_4$ photoanodes with dual-layer oxygen evolution catalysts for solar water splitting. *Science* 343, no. 6174 (February): 990–994. doi:10.1126/science.1246913.

Klahr, B., S. Gimenez, F. Fabregat-Santiago, J. Bisquert, and T. W. Hamann. 2012. Photoelectrochemical and impedance spectroscopic investigation of water oxidation with "Co–Pi"-coated hematite electrodes. *Journal of the American Chemical Society* 134, no. 40 (September): 16693–16700. doi:10.1021/ja306427f.

Krauter, S. C. W. 2007. *Solar Electric Power Generation—Photovoltaic Energy Systems: Modeling of Optical and Thermal Performance, Electrical Yield, Energy Balance, Effect on Reduction of Greenhouse Gas Emissions*. NewYork: Springer Science & Business Media.

Kuang, Y., Q. Jia, H. Nishiyama, T. Yamada, A. Kudo, and K. Domen. 2016. *Advanced Energy Materials* 6: 1501645. doi:10.1002/aenm.201670010.

Kudo, A., and Y. Miseki. 2009. Heterogeneous photocatalyst materials for water splitting. *Chemical Society Reviews* 38, no. 1 (November): 253–278. doi:10.1039/B800489G.

Lehman, P. A., and C. E. Chamberlin. 1991. Design of a photovoltaic-hydrogen-fuel cell energy system. *International Journal of Hydrogen Energy* 16, no. 5: 349–352. doi:10.1016/0360-3199(91)90172-F.

Lewis, N. S., and D. G. Nocera. 2006. Powering the planet: chemical challenges in solar energy utilization. *Proceedings of the National Academy of Sciences* 103, no. 3 (October): 15729–15735. doi:10.1073/pnas.0603395103.

Li, C., T. Hisatomi, O. Watanabe, et al. 2015. Positive onset potential and stability of Cu_2O-based photocathodes in water splitting by atomic layer deposition of a Ga_2O_3 buffer layer. *Energy & Environmental Science* 8, no. 5 (May): 1493–1500. doi:10.1039/C5EE00250H.

Li, Z., W. Luo, M. Zhang, J. Feng, and Z. Zou. 2013. Photoelectrochemical cells for solar hydrogen production: current state of promising photoelectrodes, methods to improve their properties, and outlook. *Energy & Environmental Science* 6, no. 2 (February): 347–370. doi:10.1039/C2EE22618A.

Liao, L., Q. Zhang, Z. Su, et al. 2014. Efficient solar water-splitting using a nanocrystalline CoO photocatalyst. *Nature Nanotechnology* 9 (December): 69–73. doi:10.1038/nnano.2013.272.

Lichterman, M. F., A. I. Carim, M. T. McDowell, et al. 2014. Stabilization of n-cadmium telluride photoanodes for water oxidation to O_2(g) in aqueous alkaline electrolytes using amorphous TiO_2 films formed by atomic-layer deposition. *Energy & Environmental Science* 7, no. 10 (August): 3334–3337. doi:10.1039/C4EE01914H.

Lin, C.-Y., Y.-H. Lai, D. Mersch, and E. Reisner. 2012. Cu_2O | NiOx nanocomposite as an inexpensive photocathode in photoelectrochemical water splitting. *Chemical Science* 3, no. 12 (October): 3482–3487. doi:10.1039/C2SC20874A.

Liu, J., Y. Liu, N. Liu, et al. 2015a. Water splitting. Metal-free efficient photocatalyst for stable visible water splitting via a two-electron pathway. *Science* 347 no. 6225 (February): 970–974. doi:10.1126/science.aaa3145.

Liu, X., S. K. Hoekman, C. Robbins, and P. Ross. 2015b. Lifecycle climate impacts and economic performance of commercial-scale solar PV systems: A study of PV systems at Nevada's Desert Research Institute (DRI). *Solar Energy* 119 (September): 561–572. doi:10.1016/j.solener.2015.05.001.

Luque, A., and S. Hegedus. 2011. *Handbook of Photovoltaic Science and Engineering*. New York: John Wiley & Sons.

Ma, S. S. K., K. Maeda, R. Abe, and K. Domen. 2012. Visible-light-driven nonsacrificial water oxidation over tungsten trioxide powder modified with two different cocatalysts. *Energy & Environmental Science* 5, no. 8 (August): 8390–8397. doi:10.1039/C2EE21801A.

Mahmoud, M. A., W. Qian, and M. A. El-Sayed. 2011. Following charge separation on the nanoscale in Cu_2O–Au nanoframe hollow nanoparticles. *Nano Letters* 11, no. 8 (July): 3285–3289. doi:10.1021/nl201642r.

Markvart, T. 2000. *Solar Electricity*. New York: John Wiley & Sons.

Martin, D. J., K. Qiu, S. A. Shevlin, et al. 2014a. Highly efficient photocatalytic H_2 evolution from water using visible light and structure-controlled graphitic carbon nitride. *Angewandte Chemie International Edition* 53, no. 35 (August): 9240–9245. doi:10.1002/anie.201403375.

Martin, D. J., N. Umezawa, X. Chen, J. Ye, and J. Tang. 2013. Facet engineered Ag_3PO_4 for efficient water photooxidation. *Energy & Environmental Science* 6, no. 11 (September): 3380–3386. doi:10.1039/C3EE42260G.

Martin, D. J., P. J. T. Reardon, S. J. Moniz, and J. Tang. 2014b. Visible light-driven pure water splitting by a nature-inspired organic semiconductor-based system. *Journal of the American Chemical Society* 136, no. 36: 12568–12571. doi:10.1021/ja506386e.

Mason, J., and K. Zweibel. 2008. Chapter 9. *Solar Hydrogen Generation: Toward a Renewable Energy Future*, eds. R. McConnell and K. Rajeshwar. Springer, 2008, vol. 1, 273–313. doi:10.1007/978-0-387-72810-0.

McDowell, M. T., M. F. Lichterman, A. I. Carim, et al. 2015. The influence of structure and processing on the behavior of TiO_2 protective layers for stabilization of n-Si/TiO_2/Ni photoanodes for water oxidation. *ACS Applied Materials & Interfaces* 7, no. 28 (June): 15189–15199. doi:10.1021/acsami.5b00379.

Memming, R. 2008. *Semiconductor Electrochemistry*. New York: John Wiley & Sons.

Millet, P., and S. Grigoriev. 2013. Chapter 2—Water electrolysis technologies. *Renewable Hydrogen Technologies: Production, Purification, Storage, Applications and Safety*, 19–41. doi:10.1016/B978-0-444-56352-1.00002-7.

Moniz, S. J. A., S. A. Shevlin, and D. J. Martin. 2015. Visible-light driven heterojunction photocatalysts for water splitting—A critical review. *Energy Environmental Science* 8, no. 3 (January): 731–759. doi:10.1039/C4EE03271C.

Morales-Guio, C. G., L. Liardet, M. T. Mayer, S. D. Tilley, M. Grätzel, and X. Hu. 2015. Photoelectrochemical hydrogen production in alkaline solutions using Cu_2O coated with earth-abundant hydrogen evolution catalysts. *Angewandte Chemie International Edition* 54, no. 2 (January): 664–667. doi:10.1002/anie.201410569.

Morales-Guio, C. G., S. D. Tilley, H. Vrubel, M. Grätzel, and X. Hu. 2014. Hydrogen evolution from a copper(I) oxide photocathode coated with an amorphous molybdenum sulphide catalyst. *Nature Communications* 5 (January): 3059. doi:10.1038/ncomms4059.

Muller-Langer, F., E. Tzimas, M. Kaltschmitd, and S. Peteves. 2007. Techno-economic assessment of hydrogen production processes for the hydrogen economy for the short and medium term. *International Journal of Hydrogen Energy* 32, no. 16 (November): 3797–3810. doi:10.1016/j.ijhydene.2007.05.027.

Mundada, A. S., K. K. Shah, and J. M. Pearce. 2016. Levelized cost of electricity for solar photovoltaic, battery and cogen hybrid systems. *Renewable and Sustainable Energy Reviews* 57 (May): 692–703. doi:10.1016/j.rser.2015.12.084.

Murphy, A., P. Barnes, L. Randeniya, et al. 2006. Efficiency of solar water splitting using semiconductor electrodes. *International Journal of Hydrogen Energy* 31, no. 14 (November): 1999–2017. doi:10.1016/j.ijhydene.2006.01.014.

Paola, A., Z. Fabrizio, and O. Fabio. 2011. Techno-economic optimisation of hydrogen production by PV—Electrolysis: "RenHydrogen" simulation program. *International Journal of Hydrogen Energy* 36, no. 2 (January): 1371–1381. doi:10.1016/j.ijhydene.2010.10.068.

Paracchino, A., V. Laporte, K. Sivula, M. Grätzel, and E. Thimsen. 2011. Highly active oxide photocathode for photoelectrochemical water reduction. *Nature Materials* 10 (May): 456–461. doi:10.1038/nmat3017.

Paracchino, A., N. Mathews, T. Hisatomi, M. Stefik, S. D. Tilley, and M. Grätzel. 2012. Ultrathin films on copper(I) oxide water splitting photocathodes: a study on performance and stability. *Energy & Environmental Science* 5, no. 9 (July): 8673–8681. doi:10.1039/C2EE22063F.

Paracchino, A., V. Laporte, K. Sivula, M. Grätzel, and E. Thimsen. 2011. Highly active oxide photocathode for photoelectrochemical water reduction. *Nature Materials* 10 (May): 456–461. doi:10.1038/nmat3017.

Pareek, A., P. Paik, and P. H. Borse. 2014a. Nanoniobia modification of CdS photoanode for an efficient and stable photoelectrochemical cell. *Langmuir* 30, no. 51 (December): 15540–15549. doi:10.1021/la503713t.

Pareek, A., R. Purbia, P. Paik, N. Y. Hebalkar, H. G. Kim, and P. H. Borse. 2014b. Stabilizing effect in nano-titania functionalized CdS photoanode for sustained hydrogen generation. *International Journal of Hydrogen Energy* 39, no. 9 (March): 4170–4180. doi:10.1016/j.ijhydene.2013.12.185.

Parrado, C., A. Girard, F. Simon, and E. Fuentealba. 2016. 2050 LCOE (Levelized Cost of Energy) projection for a hybrid PV (photovoltaic)-CSP (concentrated solar power) plant in the Atacama Desert, Chile. *Energy* 94 (January): 422–430. doi:10.1016/j.energy.2015.11.015.

Paul, B., and J. Andrews. 2008. Optimal coupling of PV arrays to PEM electrolysers in solar–hydrogen systems for remote area power supply. *International Journal of Hydrogen Energy* 33, no. 2 (January): 490–498. doi:10.1016/j.ijhydene.2007.10.040.

Petrakopoulou, F., J. Sanz-Bermejo, J. Dufour, and M. Romero. 2016. Exergetic analysis of hybrid power plants with biomass and photovoltaics coupled with a solid-oxide electrolysis system. *Energy* 94 (January): 304–315. doi:10.1016/j.energy.2015.10.118.

Pijpers, J. J. H., M. T. Winkler, Y. Surendranath, T. Buonassisi, and D. G. Nocera. 2011. Light-induced water oxidation at silicon electrodes functionalized with a cobalt oxygen-evolving catalyst. *Proceedings of the National Academy of Sciences* 108, no. 25 (June): 10056–10061. doi:10.1073/pnas.1106545108.

Pilli, S. K., T. E. Furtak, L. D. Brown, T. G. Deutsch, J. A. Turner, and A. M. Herring. 2011. Cobalt-phosphate (Co-Pi) catalyst modified Mo-doped $BiVO_4$ photoelectrodes for solar water oxidation. *Energy & Environmental Science* 4, no. 12 (December): 5028–5034. doi:10.1039/C1EE02444B.

Pinaud, B. A., J. D. Benck, L. C. Seitz, et al. 2013. Technical and economic feasibility of centralized facilities for solar hydrogen production via photocatalysis and photoelectrochemistry. *Energy & Environmental Science* 6, no. 7 (June): 1983–2002. doi:10.1039/C3EE40831K.

Rabinovich, E., and G. Hodes. 2013. Effective bandgap lowering of CdS deposited by successive ionic layer adsorption and reaction. *The Journal of Physical Chemistry C* 117, no. 4 (January): 1611–1620. doi:10.1021/jp3105453.

Radi, A., D. Pradhan, Y. Sohn, and K. T. Leung. 2010. Nanoscale shape and size control of cubic, cuboctahedral, and octahedral Cu–Cu_2O core–shell nanoparticles on Si(100) by one-step, templateless, capping-agent-free electrodeposition. *ACS Nano* 4, no. 3 (February): 1553–1560. doi:10.1021/nn100023h.

Rashid, M. M., M. K. Al Mesfer, H. Naseem, and M. Danish. 2015. Hydrogen production by water electrolysis: a review of alkaline water electrolysis, PEM water electrolysis and high temperature water electrolysis. *International Journal of Engineering and Advanced Technology* 4, no. 3 (February): 80–93.

Reber, J. F., and M. Rusek. 1986. Photochemical hydrogen production with platinized suspensions of cadmium sulfide and cadmium zinc sulfide modified by silver sulfide. *The Journal of Physical Chemistry* 90, no. 5 (February): 824–834. doi: 10.1021/j100277a024.

Reece, S. Y., J. A. Hamel, K. Sung, et al. 2011. Wireless solar water splitting using silicon-based semiconductors and earth-abundant catalysts. *Science* 334, no. 6056 (November): 645–648. doi:10.1126/science.1209816.

Riha, S. C., B. M. Klahr, E. C. Tyo, et al. 2013. Atomic layer deposition of a submonolayer catalyst for the enhanced photoelectrochemical performance of water oxidation with hematite. *ACS Nano* 7, no. 3 (February): 2396–2405. doi:10.1021/nn305639z.

Rocheleau, R. E., and E. Miller. 1997. Photoelectrochemical production of hydrogen: Engineering loss analysis. *International Journal of Hydrogen Energy* 22, no. 8 (August): 771–782. doi:10.1016/S0360-3199(96)00221-2.

Ross, R. T., and T. L. Hsiao. 1977. Limits on the yield of photochemical solar energy conversion. *Journal of Applied Physics* 48: 4783. doi:10.1063/1.323494.

Sakai, S., Y. Ueta, and Y. Terauchi. 1993. Band gap energy and band lineup of III-V alloy semiconducters incorparating nitrogen and boron. *Japanese Journal of Applied Physics* 32: 4413–4417. http://iopscience.iop.org/article/10.1143/JJAP.32.4413/meta.

Seabold, J. A., and K.-S. Choi. 2011. Effect of a cobalt-based oxygen evolution catalyst on the stability and the selectivity of photo-oxidation reactions of a WO_3 photoanode. *Chemistry of Materials* 23, no. 5 (March): 1105–1112. doi:10.1021/cm1019469.

Seitz, L. C., Z. Chen, A. J. Forman, B. A. Pinaud, J. D. Benck, and T. F. Jaramillo. 2014. Modeling practical performance limits of photoelectrochemical water splitting based on the current state of materials research. *ChemSusChem* 7, no. 5 (May): 1372–1385. doi:10.1002/cssc.201301030.

Shaner, M. R., S. Hu, K. Sun, and N. S. Lewis. 2015. Stabilization of Si microwire arrays for solar-driven H_2O oxidation to O_2(g) in 1.0 M KOH(aq) using conformal coatings of amorphous TiO_2. *Energy & Environmental Science* 8, no. 1 (January): 203–207. doi:10.1039/C4EE03012E.

Shapiro, D., J. Duffy, M. Kimble, and M. Pien. 2005. Solar-powered regenerative PEM electrolyzer/fuel cell system. *Solar Energy* 79, no. 5 (November): 544–550. doi:10.1016/j.solener.2004.10.013.

Shell, C., T. Photoanodes, R. H. Coridan, et al. 2014. Photoelectrochemical behavior of hierarchically structured Si/WO_3 core–shell tandem photoanodes. *Nano Letters* 14, no. 5 (March): 2310–2317. doi:10.1021/nl404623t.

Shi, M., X. Pan, W. Qiu, D. Zheng, M. Xu, and H. Chen. 2011. Si/ZnO core–shell nanowire arrays for photoelectrochemical water splitting. *International Journal of Hydrogen Energy* 36, no. 23 (November): 15153–15159. doi:10.1016/j.ijhydene.2011.07.145.

Shi, X., I. Y. Choi, K. Zhang, et al. 2014. Efficient photoelectrochemical hydrogen production from bismuth vanadate-decorated tungsten trioxide helix nanostructures. *Nat Commun* 5 (September): 4775. doi:10.1038/ncomms5775.

Spasiano, D., R. Marotta, S. Malato, P. Fernandez-Ibañez, and I. Di Somma. 2015. Solar photocatalysis: Materials, reactors, some commercial, and pre-industrialized applications. a comprehensive approach. *Applied Catalysis B: Environmental* 170–171 (July): 90–123. doi:10.1016/j.apcatb.2014.12.050.

Steeb, H., A. Mehrmann, W. Seeger, and W. Schnurnberger. 1985. Solar hydrogen production: Photovoltaic/electrolyzer system with active power conditioning. *International Journal of Hydrogen Energy* 10, no. 6: 353–358. doi:10.1016/0360-3199(85)90060-6.

Steinmiller, E. M. P., and K.-S. Choi. 2009. Photochemical deposition of cobalt-based oxygen evolving catalyst on a semiconductor photoanode for solar oxygen production. *Proceedings of the National Academy of Sciences of the United States of America* 106, no. 49 (October): 20633–20636. doi:10.1073/pnas.0910203106.

Sun, K., F. H. Saadi, M. F. Lichterman, et al. 2015a. Stable solar-driven oxidation of water by semiconducting photoanodes protected by transparent catalytic nickel oxide films. *Proceedings of the National Academy of Sciences* 112, no. 12 (February): 3612–3617. doi:10.1073/pnas.1423034112.

Sun, K., M. T. McDowell, A. C. Nielander, et al. 2015b. Stable solar-driven water oxidation to O_2(g) by Ni-oxide-coated silicon photoanodes. *The Journal of Physical Chemistry Letters* 2: 592–598. doi:10.1021/jz5026195.

Sun, K., S. Shen, Y. Liang, P. E. Burrows, S. S. Mao, and D. Wang. 2014. Enabling silicon for solar-fuel production. *Chemical reviews* 114, no. 17 (August): 8662–8719. doi:10.1021/cr300459q.

Sun, K., Y. Kuang, E. Verlage, B. S. Brunschwig, C. W. Tu, and N. S. Lewis. 2015c. Functional coatings: sputtered NiOx films for stabilization of p+n-InP photoanodes for solar-driven water oxidation (Adv. Energy Mater. 11/2015). *Advanced Energy Materials* 5, no. 11 (June): 1402276. doi:10.1002/aenm.201570059.

Surendranath, Y., D. A. Lutterman, Y. Liu, and D. G. Nocera. 2012. Nucleation, growth, and repair of a cobalt-based oxygen evolving catalyst. *Journal of the American Chemical Society* 134, no. 14 (March): 6326–6336. doi:10.1021/ja3000084.

Tanaka, N. 2008. Energy technology perspective: Scenario and strategies to 2050. Paris: IEA. http://www.iea.org/media/etp/etp2008.pdf.

Tanaka, N. 2010. Energy technology perspectives: scenarios and strategies to 2050. Paris: IEA. https://www.iea.org/publications/freepublications/publication/etp2010.pdf.

Tidball, R., J. Bluestein, N. Rodriguez, and S. Knoke. 2010. Cost and performance assumptions for modeling electricity generation technologies. National Renewable Energy Laboratory. doi:10.2172/993653.

Tran, P. D., S. K. Batabyal, S. S. Pramana, J. Barber, L. H. Wong, and S. C. J. Loo. 2012. A cuprous oxide–reduced graphene oxide (Cu_2O–rGO) composite photocatalyst for hydrogen generation: employing rGO as an electron acceptor to enhance the photocatalytic activity and stability of Cu_2O. *Nanoscale* 4, no. 13 (July): 3875–3878. doi:10.1039/C2NR30881A.

Turner, J. A. 2014. Shining a light on solar water splitting—Response. *Science* 344: 469–470. doi:10.1126/science.344.6183.469-b.

van de Krol, R. 2012. Principles of photoelectrochemical cells. *Photoelectrochemical Hydrogen Production*, ed., R. van de Krol, and M. Grätzel. Boston, Springer, 13–67.

van Dorp, D. H., N. Hijnen, M. Di Vece, and J. J. Kelly. 2009. SiC: A photocathode for water splitting and hydrogen storage. *Angewandte Chemie International Edition* 48, no. 33 (August): 6085–6088. doi:10.1002/anie.200900796.

Villa, K., X. Domènech, S. Malato, M. I. Maldonado, and J. Peral. 2013. Heterogeneous photocatalytic hydrogen generation in a solar pilot plant. *International Journal of Hydrogen Energy* 38, no. 29 (September): 12718–12724. doi:10.1016/j.ijhydene.2013.07.046.

Walter, M. G., E. L. Warren, J. R. McKone, et al. 2010. Solar water splitting cells. *Chemical Reviews* 110, no. 11 (November): 6446–6473. doi:10.1021/cr1002326.

Wang, L., F. Dionigi, N. T. Nguyen, et al. 2015. Tantalum nitride nanorod arrays: introducing Ni–Fe layered double hydroxides as a cocatalyst strongly stabilizing photoanodes in water splitting. *Chemistry of Materials* 27, no. 7 (March): 2360–2366. doi:10.1021/cm503887t.

Wang, M., L. Sun, Z. Lin, J. Cai, K. Xie, and C. Lin. 2013. p–n Heterojunction photoelectrodes composed of Cu_2O-loaded TiO_2 nanotube arrays with enhanced photoelectrochemical and photoelectrocatalytic activities, *Energy & Environmental Science* 6, no. 4 (April): 1211–1220. doi:10.1039/C3EE24162A.

Wang, Q., T. Hisatomi, and Q. Jia, 2016 Scalable water splitting on particulate photocatalyst sheets with a solar-to-hydrogen energy conversion efficiency exceeding 1%, *Nature Materials* 15 (March): 611–615. doi:10.1038/nmat4589.

Wang, Q., Y. Xie, F. Soltani-Kordshuli, and M. Eslamian. 2016. Progress in emerging solution-processed thin film solar cells—Part I: Polymer solar cells. *Renewable and Sustainable Energy Reviews* 56 (April): 347–361. doi:10.1016/j.rser.2015.11.063.

Wang, X., K.-Q. Peng, Y. Hu, et al. 2013. Silicon/hematite core/shell nanowire array decorated with gold nanoparticles for unbiased solar water oxidation. *Nano Letters* 14, no. 1 (January): 18–23. doi:10.1021/nl402205f.

Weber, M. F., and M. J. Dignam. 1984. Efficiency of splitting water with semiconducting photoelectrodes. *Journal of The Electrochemical Society* 131, no. 6: 1258. doi: 10.1149/1.2115797.

Weber, M. F., and M. J. Dignam. 1986. Splitting water with semiconducting photoelectrodes—Efficiency considerations. *International Journal of Hydrogen Energy* 11, no. 4: 225–232. doi:10.1016/0360-3199(86)90183-7.

Wei, J., Y. Feng, P. Zhou, et al. 2015. A bioinspired molecular polyoxometalate catalyst with two cobalt(II) oxide cores for photocatalytic water oxidation. *ChemSusChem* 8, no. 16 (August): 2630–2634. doi:10.1002/cssc.201500490.

Wenham, S. R., M. A. Green, M. E. Watt. and R. Corksih. 2009. *Applied Photovoltaics*. London: Earthscan.

Wu, K., Z. Chen, H. Lv, H. Zhu, C. L. Hill, and T. Lian. 2014. Hole removal rate limits photodriven H2 generation efficiency in CdS-Pt and CdSe/CdS-Pt semiconductor nanorod–metal tip heterostructures. *Journal of the American Chemical Society* 136, no. 21 (May): 7708–7716. doi:10.1021/ja5023893.

Wu, L., L.-k. Tsui, N. Swami, and G. Zangari. 2010. Photoelectrochemical stability of electrodeposited Cu_2O films. *The Journal of Physical Chemistry C* 114, no. 26 (June): 11551–11556. doi:10.1021/jp103437y.

Xing, Z., X. Zong, J. Pan, and L. Wang. 2013. On the engineering part of solar hydrogen production from water splitting: Photoreactor design. *Chemical Engineering Science* 104 (December): 125–146. doi:10.1016/j.ces.2013.08.039.

Yu, Q., X. Meng, T. Wang, P. Li, and J. Ye. 2015. Hematite films decorated with nanostructured ferric oxyhydroxide as photoanodes for efficient and stable photoelectrochemical water splitting. *Advanced Functional Materials* 25, no. 18 (May): 2686–2692. doi:10.1002/adfm.201500383.

Zhong, D. K., S. Choi, and D. R. Gamelin. 2011. Near-complete suppression of surface recombination in solar photoelectrolysis by "Co-Pi" catalyst-modified W:$BiVO_4$. *Journal of the American Chemical Society* 133, no. 45 (September): 18370–18377. doi:10.1021/ja207348x.

Zou, X. and Y. Zhang. 2015. Noble metal-free hydrogen evolution catalysts for water splitting. *Chemical Society Reviews* 44, no. 15 (April): 5148–5180. doi:10.1039/C4CS00448E.

3

Material Selection for Photoelectrochemical or Photocatalytic Processes

Balasubramanian Viswanathan

CONTENTS

3.1 Introduction

Hydrogen economy has been at our doorsteps for a few decades now. However, the three arms of this economy, namely, production of hydrogen, storage, and distribution, have not reached any matured and possible commercial stage. Hydrogen production and storage are still in the domain of scientific research and development. Among the various options available (briefly outlined in Table 3.1) for the production of hydrogen, photoelectrochemical (PEC) or photocatalytic production of hydrogen by the decomposition of water is considered as one of the viable routes. This is a self-imposed restriction, and possibly some unique advantages can be claimed such as the possible equidistribution of energy source among the various nations since the source of energy in this case, the solar radiation that is supposed to be common among all nations though the extent of solar radiation available to each of these nations, may differ.

The basis of PEC decomposition of water can be considered equivalent to electrolysis of water wherein instead of the electrochemical potential, photon energy (or photon energy and bias voltage) is applied to surmount the normally uphill process of water decomposition. In essence, the normally positive free energy changes of reactions of hydrogen and oxygen evolution are adjusted to negative values by the photons, and thus the reducing electron combining with H^+ ions becomes free energy change negative due to excitation of electrons from the valence band to the conduction band. Similarly, the

TABLE 3.1

Some of the Possible Sources for Hydrogen

1. Steam reforming of naphtha and natural gas
2. Chemical decomposition of water
3. Electrolysis of water
4. Thermochemical routes for the decomposition of water
5. Photoelectrochemical (photocatalytic) decomposition of water
6. Biochemical routes for the decomposition of water

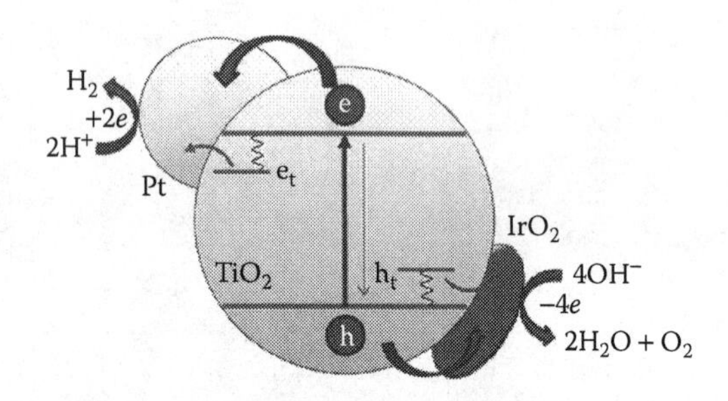

FIGURE 3.1
Pictorial representation of water decomposition on a TiO_2 semiconductor by photons of energy greater than E_g; e_t and h_t are electron and hole traps, respectively.

potential of oxidizing species, namely, the holes generated due to photoexcitation are suitable for oxygen evolution. There can be additional electron/hole transfer agents such as Pt or IrO_2, and this situation is pictorially shown in Figure 3.1.

Thus, the basic process is driven by the photoinduced charge transfers occurring on the irradiated semiconductor surface with the initiation of a variety of redox conversion reactions. Therefore, the main aim of this exercise is to find appropriate materials to harness solar energy effectively and in the desired reaction. This means that the material chosen should satisfy some minimum conditions. They can be listed as follows:

1. The band gap of the semiconducting material chosen should be such that the maximum solar radiation is absorbed.
2. The absorbance of solar radiation by the semiconductor should be high.
3. The bottom of the conduction band and the top of the valence band should be at such energy positions so that the reduction and oxidation reactions are both downhill (in this case, the hydrogen evolution potential that is assumed to be zero in the electrochemical scale and hence the conduction band bottom should be more negative with respect to zero in the electrochemical scale) and for the oxidation reaction vice versa.

If both these conditions are not satisfied, then water decomposition cannot be complete. If the bottom of the conduction band is more positive to the hydrogen evolution potential, then this reaction will not take place. Similarly, if the top of the valence band is more positive to the oxygen evolution potential, then this reaction is not feasible. When both bottom of the conduction band and top of the valence band saddled between hydrogen and oxygen evolution potentials, then that system cannot decompose water. These situations are pictorially shown in Figure 3.2. Nanoscale photocatalysts function in the decomposition of water in a similar manner. The advantages of these systems are as follows:

- High surface area (promotes charge-transfer processes at the solid–liquid interface and thus only low overpotentials are required; this in turn reduces the need for expensive cocatalysts)
- Short carrier pathways (ideal electron and hole collection is possible when particle diameter (d) is less than the mean free diffusion length of electrons and holes, L_e and L_h)
- Improved light management (light absorption/scattering is maximum)
- Quantum size confinement (increases the thermodynamic driving force for interfacial charge transfer and hence the rate for interfacial charge transfer is also increased)

However, there are some restrictions on the use of nanomaterials as effective materials for water decomposition. The essential ones are (1) reduced space charge layer (charge-carrier separation is less effective) and (2)

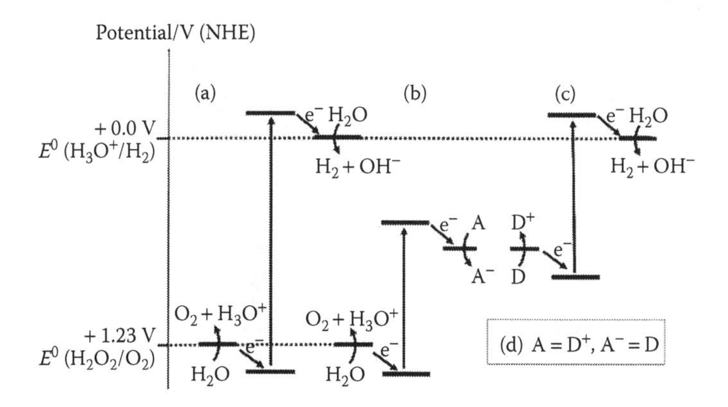

FIGURE 3.2
Potential energy diagrams for photochemical water splitting at pH 0: (a) single semiconductor system, (b) with an electron acceptor, (c) with an electron donor, and (d) dual semiconductor system (z scheme) employing a redox shuttle. (Reprinted with permission from Osterloh, F. E., *Chemistry of Materials*, 20, 35–54, 2008. Copyright [2008] the American Chemical Society.)

increased surface recombination (surface and interfacial recombination rates are enhanced due to high interface area). In the case of material selection, one has to focus on the light absorption characteristics of the material since the water decomposition and the efficiency of the process depend on the light absorption characteristics of the material. The various possibilities of the nanostate of catalysts in the case of light absorption are shown in Figure 3.3. The light penetration length is dependent on the geometry of the light absorbing material. The shape and size of the light absorbing particles are particularly important. The energy positions of the conduction band bottom and that of the valence band maximum will also depend on the size of the particle. Since these two energetic positions are important for the water decomposition reaction, their alterations are pictorially shown in Figure 3.3b as a function of the size of the particle.

There are various semiconductor catalyst systems that have been tried for the photocatalytic decomposition of water. Selected list of combination of semiconductors that have yielded satisfactory results are assembled in Table 3.2. The data given in Table 3.2 are the best choice materials from the recent literature for the photocatalytic decomposition of water but still do not lead to postulate principles for the selection of materials. The choice of materials for PEC applications has thus not yet resulted in the desired result so as to be able to *a priori* predict the best choice of material. Several criteria have been proposed and tested such as the value of band gap, position of the

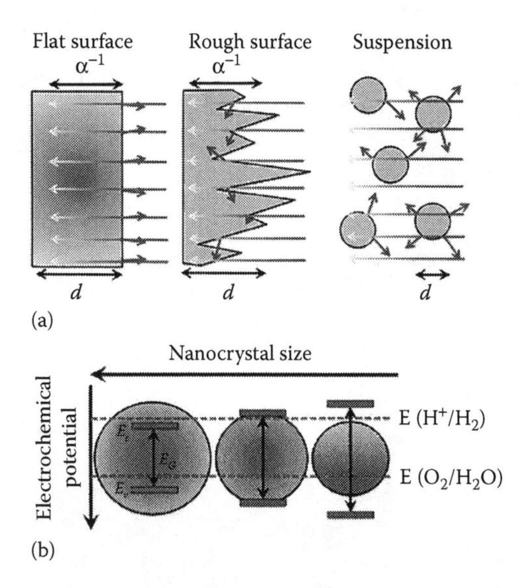

FIGURE 3.3
The light scattering from various types of surfaces (a) and the band gap variation as a function of the size of catalyst particle (b). (Osterloh, F. E., *Chemical Society Review*, 42, 2294–2320, 2013. Reproduced by permission of the Royal Society of Chemistry.)

TABLE 3.2

Selected Photocatalytic Systems That Have Been Tried for Overall Water Decomposition

Photocatalyst	Hydrogen Yield (μmol/h)	% RuO_2	Band Gap (eV)	Surface Area (m^2/g)	AQY (%)	Reference
NiO-$NaTaO_3$	2000	0	4.0	34	56 (270 nm)	Kato et al. (2003)
NiO-$Ln_2Ti_2O_7$	8000 (20 h)	0	3.82	~1	NA	Hwang et al. (2003)
RuO_2-$CaIn_2O_4$	50 (4 h)	1	~3.2	3.3	NA	Sato et al. (2003)
Ga and La doped SrTiO3	~9 mmol	0	3.2	NA	NA	Takata and Domen (2009)
RuO_2-$LiNbO_3$	47	1	4.0	28	0.7 (254 nm)	Saito et al. (2011)
RuO_2-Zn_2GeO_4	17.4	3	4.5	14.8	NA	Yan et al. (2011)
NiO_x-$SrTiO_3$	19.4	0	3.3	NA	0.041 (315 nm)	Townsend et al. (2012)
RuO_2-CdS-Ti-MCM-48	120	1	2.6	797	1.5 (550 nm)	Peng et al. (2013)

valence band and conduction band edges, and ionic character of the bonds in the semiconductor, and yet none of the postulates has the capacity to correctly predict or propose the best choice of material.

Among the various systems studied, TiO_2 occupies a prominent place in PEC and photocatalytic applications. Though anatase is the preferred phase, there is still controversy over the two polymorphs of TiO_2 namely anatase and rutile. This situation is particularly interesting since many geometrical arrangements like core shell have been proposed for the active phase. In this case of coupled semiconductor systems, the question is which are the reducing or oxidizing centers? For example, the reducing center depends on the energy position of the bottom of the conduction band with respect to the hydrogen evolution potential in the electrochemical scale. In a detailed recent investigation (Scanlon et al. 2013), this aspect has been addressed. It is generally believed based on impedance measurements that the conduction band in anatase is 0.2 eV more negative with respect to that of rutile (see Figure 3.4), which means if photoexcitations were to occur in anatase, the reduction reaction will take place on rutile. Alternatively, recent photoemission measurements showed that the work function of rutile is 0.2 eV less than that of anatase, thus placing the conduction band bottom is 0.2 eV more negative with respect to the conduction band bottom of anatase. This investigation has thrown open the reduction/oxidation reactions taking place in polymorphic semiconductors when they coexist in doubt.

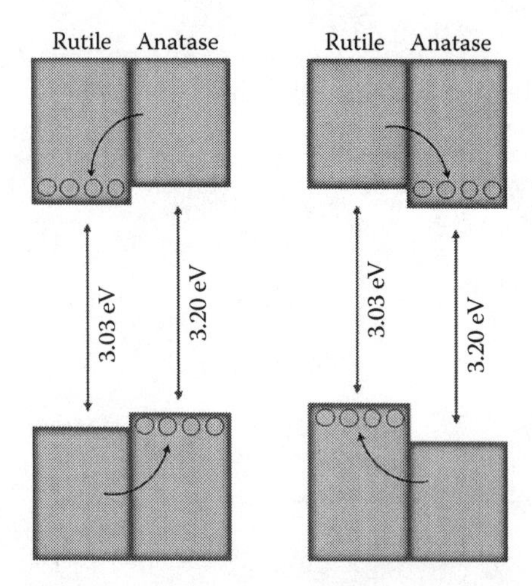

FIGURE 3.4
Two proposed valence and conduction band alignments for the anatase/rutile interface.

Similar concepts have been used to explain the behavior of coupled semiconductor systems or sensitized (dye or other molecular systems) (de Angelis et al. 2008; Jeyalakshmi et al. 2016) configurations without concern of the variations or interpenetration of the two phases at the interface. This situation is particularly interesting that many studies proposed sensitization as one of the means of achieving the necessary efficiency for the process. Energy levels will vary smoothly at the interface, though in isolated systems the occupied and unoccupied states may have discrete values. How do these band bendings at the interface facilitate the smooth electron transfer has been the subject of many investigations in the past. It is presumed that smooth interpenetration of multiple phases gradually alters the energy position of electron and hole states at the interface (this situation holds good even for semiconductor vacuum interface or semiconductor–air interface) and thus helps in surmounting the energy barrier for charge transfer. This aspect has to have clarity, and precise predictive capability will evolve from future studies in this direction.

It is therefore clear that the band alignments only account for the thermodynamic control of the process; for the control of the kinetic aspects of the process, one has to seek the solution elsewhere and this aspect has not yet been fully addressed in the literature (Kalyanasundram and Gratzel 2010).

The situation is pictorially shown in Figure 3.5. The thermodynamic energetics of the charge carriers in the coupled system is shown in Figure 3.5a. Even though the energetics may be favorable, the kinetics of recombination will disfavor the reaction. Therefore, kinetically the lifetime of the charge carriers has to be improved, and this process must be facile in the coupled systems. The charge carriers should not have many alternate deexcitation routes, while the charge transfer to water to decompose must be the preferred route.

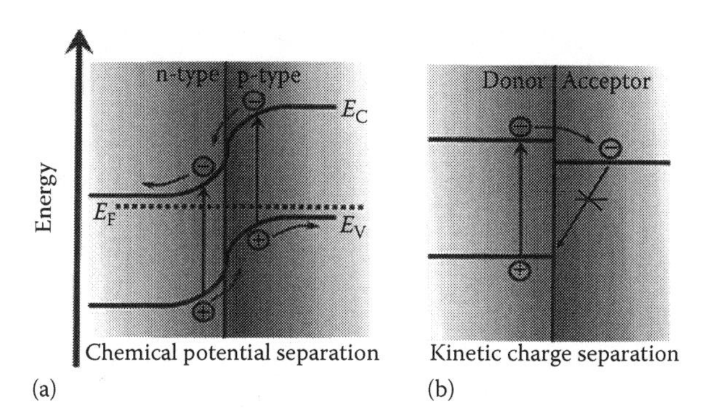

FIGURE 3.5
Energy-level positions in coupled semiconductors in thermodynamic (a) and kinetic control, respectively (b).

The band alignment in the polymorphs of TiO_2 is still a question. The highly photocatalytic active form is its mixed phase (80% anatase and 20% rutile the so-called P-25) of two polymorphs of anatase and rutile rather than their pristine compositions. Such a synergetic effect is understood by the staggered band alignment of conduction and valence band edges of the two phases favorable to spatial charge separation. However, electron migration in either direction between the two phases has been reported, meaning the band positions of the conduction band bottom edges of the two phases are at the same energy level.

If the conduction band bottom level in both the polymorphs is at the same energy level, electron migration in both directions is feasible and hence the reduction reaction is feasible whichever phase is exposed on the surface, while oxidation reaction is feasible only on the rutile phase. This postulate possibly provides clues why the photocatalytic activity is dependent on the particle sizes of the polymorphs. The latest study shows that the band alignments and the direction of charge transfer are still unknown, and the clarity on this issue alone can provide predictive capability of materials for the PEC decomposition of water (Mi and Weng 2015). In an extensive photospectroscopic study, it was shown that the top of the valence band of rutile is ~0.7 eV less positive with respect to that of anatase (Kullgren et al. 2015).

The relative positions of bottom of the conduction band (reducing power) and top of the valence band (oxidizing power) of the material are possibly controlled by the ionicity of the bond between the cation and anion of the semiconductor. This aspect has to be thoroughly investigated, and appropriate percentage ionicity of the bond has to be identified suitable for facile charge transfer at the semiconductor electrolyte interface. But still it is not certain if this will provide the answer to this problem of predicting suitable catalyst systems for the facile PEC decomposition of water.

The positions of top of the valence band and bottom of the conduction band are necessary conditions for the smooth reduction and oxidation reactions in general and in particular the water decomposition. Normally if in a semiconductor, the band positions were not suitable for water decomposition, then particle size effect can be used to alter the energy positions of valence band maximum and conduction band minimum so as to be suitable for water decomposition. This situation is pictorially shown as quantum size effect of the semiconductor in Figure 3.3. Alternatively, the presence of potential determining ions and its surrounding effect can also be used to alter the position of the valence band maximum (oxidizing power) and conduction band minimum (reducing power), and this situation is pictorially shown in Figure 3.6. Thus, PEC decomposition of water has been considered as one of the viable options for obtaining hydrogen fuel. Intense research has gone on for the past five decades to make these process economically viable. In spite of these concerted attempts, there appears to be some issues that are yet to be resolved with respect to improving the efficiency of the process. Over 360 different semiconducting materials and the possible variations in them such as doping, coupling, homo- and heterojunctions, composite, multilayer, and complex systems among them have all been tried out, but still the efficiency of the photoelectrolysis has not reached the desired levels, (Osterloh 2008).

This situation has been analyzed at various levels, and it has even been considered whether the process of PEC decomposition is a dream or reality (Viswanathan 2005). The situation has also been reviewed recently (Chen et al. 2010; Minggu et al. 2010). Various arguments have also been given possibly to understand and remedy the situation. These reasons have led to the development of methodology known as band gap engineering, which is essentially trying to alter the energy states of the valence and conduction band edges of the semiconductors employed. This situation may remain so for some more time in spite of the earnest efforts of the scientists to remedy it.

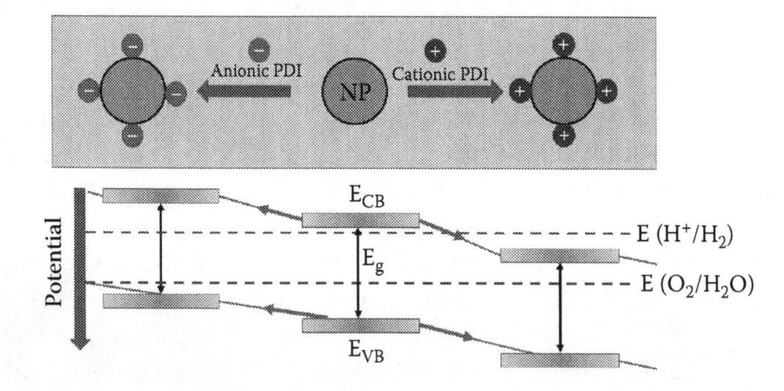

FIGURE 3.6
The alteration of the band positions by the influence of potential determining ions. (Osterloh, F. E., *Chemical Society Review*, 42, 2294–2320, 2013. Reproduced by permission of the Royal Society of Chemistry.)

3.2 Study of the Interface

One of the fundamental issues in understanding of this process is the details of the interface structure in the presence of both electric and photon fields. The electrode/electrolyte interface has been examined at various stages of development of science and has been critically considered recently (Damaskin and Petrii 2011). Though various models have been periodically proposed to detail the electrode/electrolyte interface, the description is still incomplete. One has to make various kinds of assumptions such as no adsorption on electrode surface, or only electrostatic interaction between the electrified interface and the oriented ions or dipoles from the electrolyte side, or even considering the diffuse nature of the double layer extending over longer molecular dimensions to derive measurable parameters such as the capacitance of the double or multiple layers formed at the interface. These attempts have been made with respect to the metal/electrolyte interface. Our understanding of the metal/electrolyte interface itself is incomplete where the field penetration in the solid phase is only skin depth and hence always neglected (Viswanathan and Scibioh 2009). In the case of semiconductors, which are employed as the photochemical sensors for PEC applications, the field gradient can be present on both sides of the interphase and this can add to the problem. Any model will not ultimately be a substitute for the reality but, however, model may give us some clues on the behavior of real systems. In this presentation, one wishes to build up a model like the electrode/electrolyte model in electrochemistry and use this model to outline the behavior of the interface in the photoelectrochemical (PE) cells, which, it is hoped, will pave the way in the direction of appropriate semiconductor anode to be employed in these cells. The proposed model can essentially retain all the postulates for the electrode/electrolyte interface; in addition, it should incorporate the following aspects.

1. The interaction of the electric field with the simultaneous photon field applied. The exact changes that will occur in the occupied and unoccupied states in semiconductor are not clear but the net charge density is altered and hence the field gradient may also change. The photon field creates excitation of electrons from the valence band to the conduction band, and the conduction band wave functions possess "d" character of metal ions while the valence band wave functions have the "p" character in the case of oxide semiconductors; the net charge density is altered and hence the interaction with water molecules can be different when the photon field is applied in addition to the electric field.
2. The influence of electric field on the altered charge distribution effected in the semiconductor due to the photon field has to be considered. This may effectively alter the field gradient in the solution

side and also the field gradient in the semiconductor side. This is usually termed as the Garrett Brattain space charge region. The occupancy of the conduction band states can give rise to a different field penetration in the semiconductor as compared to field penetration without the photon field.

3. The accumulation of species (charged or otherwise) at the interface due to altered charge density in the semiconductor due to the photon field has to be taken into account. This can be followed up in terms of the wave function overlaps from the semiconductor and water molecule, and this could be different in the presence of the photon field.
4. Electric field induced adsorption is already known in the electrode/electrolyte interface, but the superimposed photon field and how this will affect the adsorption process has not been explicitly known, though it is generally believed that photon-induced adsorption (the so-called photoadsorption) may be rare, but it may not be so since the density of states and occupancy of the states are altered as a result of the photon field.
5. At least the presence of these two fields can be expected to alter the following parameters, namely, the density of states, the perturbation of the surface states, and the net adsorption capacity. All these parameters can be expected to alter the capacity of the double (structured/conceived layers) layer presumably formed.
6. Since the semiconductor can possess some defect sites (either vacancy or alter-valent ions to give rise to semiconducting property) and the nature of frontier wave functions can be different, the nature and orbital throw of these wave functions are not immediately amenable in terms of contour maps and hence assessing the interaction of them with the substrate wave functions is not possible. Generally, it is believed that the currently available theoretical calculations on such condensed matter only gives rough estimate of both the eigenvalue and eigenfunctions of the frontier orbitals. This lack of knowledge may be the stumbling block for designing suitable semiconductors for PEC applications.

3.3 Possible Model

It is to be remarked that the model proposed has not captured all the details but has considered some of the aspects of the double layer that may be present in the PE cells. Building a pictorial model at least taking the aspects considered appears to be difficult. First, let us see what consequences one will be able to experience in the semiconductor/electrolyte interface due to these considerations. When a semiconductor electrode comes into contact with an electrolyte, the band edges at the interface are bent, and this is well known and has been treated extensively in the literature. The term band bending is not appropriate; what is meant is the energy levels on both sides are distorted smoothly so that no discrete discontinuity occurs at the interface. This band bending can not only affect the successive electron or hole transfer process but can also alter the oxidizing power and reducing power of the semiconductor. In this proposal, we assume that redox values of the depolarizer do not change; this assumption is also not valid since redox values of the depolarizer are given for the free state and not at the interface. For example, the reduction potential of hydrogen ions is assumed to be zero when the activity of hydrogen ions in the electrolyte is unity. But this situation may not be holding good at the interface and the activity of the redox species could have changed as a result of interface formation, and hence one has to take this alteration into account. Let us consider that the bands in the semiconductor are bent upward, though this terminology of upward may not be the correct way of stating, what is really meant is the bands take allowed energy states in the less positive and more negative values of potential. This can certainly affect the starting redox values of the semiconductor, but will also contribute to altering the energy barrier for charge-transfer process. Leaving these consequences, one can also envisage that the position of the Fermi level in the semiconductor will change and also the density of states of the semiconductor will change. These changes will have considerable consequences in the redox process that will take place at the interface.

Let us consider the application of the photon field only on the semiconductor. This will give rise to population of the conduction band with electrons and the valence band with positive holes. This situation should alter the binding of the energy states of the allowed energy levels in these two bands, and the direction to which the binding energy will change can depend on various factors. However, this binding energy change can be expected to cause alterations in the redox values of the semiconductor. It can be assumed that the oxidation potential possibly goes to more positive values due to changes of the top of the valence band, and the reduction potential can be expected to change to more negative values as the conduction band may move upward (?) due to electron–electron repulsion. In this discussion, we have not considered the excitations possible in the depolarizer as we assume that the wavelength of the photon field does not match with the resonance excitation in the depolarizer.

This may be the situation when there is no external electric field (bias) applied, and the only electric field is the inherent one due to the interface formation. What are the consequences of this situation? (1) The relative redox values change favorably and hence one experiences increased reactivity in the case of water decomposition and one may observe increased water decomposition. (2) The reduction power of the semiconductor is altered favorably, and hence increased hydrogen evolution is observed while the valence band alterations are such that oxygen evolution is inhibited, and this may account for the observation where oxygen evolution is not in the stoichiometric value or even no oxygen evolution at all. This could arise due to the changes that may take place in the position of the top of the valence band.

It is easy to recognize all the static parameters that are responsible for the PEC decomposition of water or any other photochemical transformations that one wishes to drive. This may be one of the reasons why band gap engineering has been taken as a major step for the development of new materials for PEC applications. However, it is time for us to consider seriously whether band gap engineering (shifting the response of the semiconductor to lesser energy radiations as compared to UV irradiation required for the most often investigated photoanode of PEC, namely, TiO_2) is the correct step for the development of efficient photoanode, or should one consider how to synergize the effect of both the electric and photon fields on both the semiconductor and the depolarizer in such a way energetically that both reduction and oxidation reactions occur at the same or nearly the same rate so that the net efficiency observed will be in the economically viable range.

The consequences of these additions to the existing model are as follows:

1. It may account why some systems are preferentially promoting one of the two reactions of water decomposition. This will be the case in what way the position of the Fermi level and the positions of the bottom edge of the conduction band and top of the valence band are altered either due to the photon field or due to the interface inherent field or biased electric field. If the edges of the band positions are shifted to more or less positive values or more negative or less negative values accordingly, it will affect the oxygen and hydrogen evolution reactions, respectively.
2. Since these alterations only change the net free energy change taking place in the evolution of these two gases, it may not be directly reflected in the kinetics of gas evolution reactions.
3. The gas evolution reactions are controlled by a variety of steps such as molecule formation from atomic species, bubble formation, and gas desorption step and any of these steps may be slow, and these are not controlled by the electron transfer steps and hence evolution rate can be small.
4. Populating the conduction band by the photon field may apparently increase the semiconductor's reducing power while the oxidizing power of the semiconductor can be altered either way since the presence of positive charge in the valence band can alter the binding energy of the states of the valence band.

The foregoing arguments stressed the importance of band gap in the choice of materials for PEC applications. Catlow et al. (2010) have analyzed this problem under theoretical models. They have reviewed the developments and applications of computational modeling techniques in the field of materials for energy technologies including hydrogen production and storage, energy storage and conversion, and light absorption and emission. They have postulated how the light absorption in these systems can be altered by appropriate substitutions. If the conduction band position is appropriate for the reduction reaction, then substitution of 13–15 (e.g., Sn) group element will alter the valence band, thus reducing the band gap. For these systems, the top of the valence band is influenced by the s^2 cation, while the bottom of the conduction band is determined by the d^0 cation. All these modifications that can be termed as sensitization of the semiconductor should not affect the stability of the semiconductor under the experimental conditions employed in PEC applications. Thus, to engineer the band gap will also have direct consequences on the defect properties of a material, and hence its conductivity, and on its stability under the influence of electrolyte.

The issues on hand can be listed as follows. (1) Identify and characterize thoroughly semiconductor materials that have appropriate values of band gaps and appropriate band edge positions and are stable in aqueous solutions. (2) Study multijunction semiconductor systems for higher efficiency water splitting achieving at least at the level of Carnot efficiency. (3) Develop techniques for the energetic control of the semiconductor electrolyte interphase. The structure of the electrolyte–electrode interface and appropriate theory for the electrode/electrolyte interface are yet to be developed. (4) Develop techniques for the preparation of transparent catalytic coatings and their application to semiconductor surfaces. (5) Identify environmental factors (e.g., pH, ionic strength, and solution composition) that affect the energetics of the semiconductor, the properties of the catalysts, and the stability of the semiconductor. (6) Identify and optimize the active sites for both hydrogen and oxygen evolution sites on the two electrodes.

Based on the structures of the semiconductors that have been examined for the PEC decomposition of water, it is deduced that they can have a marked influence on the catalytic activity. Active catalysts are generally found to have efficient charge-transport pathways that connect the interior with the surface where water splitting occurs. Oxide-bridged metal ions that can assume such a role are present in the majority of structures of oxides used for the water decomposition reaction. The structures of mixed oxides that contain isolated metal oxide polyhedral show low catalytic activities. Oxides that have distortions in metal oxide polyhedra possibly show catalytic activity. The cation size of some of dopants could be correlated with the distortion of the structure and the O_2 production rate and inversely with the H_2 production rate. Cavities in the structures of the so-called tunnel structures have also been implicated in electrical contact to the cocatalyst. Because of their small size (~1 nm), the cocatalyst particles are believed to fit into the opening of the tunnels. Finally, the high activities of some of the layered perovskites are attributed to the incorporation of water into the interlayer space. Water incorporation increases the interfacial area and reduces the necessary distance for charge transfer, thus facilitating its decomposition.

Because of their physical relationship, crystal morphology and specific surface area are the variables that must be considered together. In general, an increase in the specific surface area (i.e., a reduction in the crystal size) leads to higher catalytic activities. The same applies to the layered perovskites, which have the ability to increase their interfacial area by incorporating water into the interlayer space. Finally, there are also cases where the surface area does not matter or where the activity diminishes with an increase in the surface area.

Another postulate in generalizing the activities of the systems concerns the d electron configuration of the metal ion. Only metal compounds with d^0 ions (Ti, Zr, Nb, and Ta) and d^{10} ions (Ga, In, Ge, Sn, and Sb) have activity for overall photochemical splitting of water. Oxides are dominant, but nitrides and oxynitrides are also known to catalyze the reaction. However, this generalization could not yield predictive capability.

Thus, there are a variety of attempts to generalize the postulates for predicting the suitable semiconductor systems that will give rise to facile PEC or photocatalytic decomposition of water (Osterloh and Parkinson 2011; Walter et al. 2010). But none of these generalizations have so far led us to be able to predict the most efficient and active system.

3.4 Conclusion

It has been argued (Viswanathan 2003) that in binary oxides as the ionic nature of the bond changes the valence band that is mostly consisting of 2p orbitals of oxygen and the conduction band that is mainly contributed by the empty d orbitals of the metal ions. These arguments are only reflecting the static effect on the charge density in the case of cations and anions of the oxide semiconductors employed. Though such static conditions have been well recognized in the literature and also based on these arguments the selection of materials has been proposed, it appears that the redox values of the depolarizer and also the energy levels of the semiconductor are becoming floating on the application of both the photon and electric fields. This interplay has to be clearly understood before predictions on the choice of materials can be made. This will be one of the essential steps in the prediction of materials for PEC applications.

References

Catlow, C. R. A., Z. X. Guo, M. Miskufova, et al. 2010. Advances in computational studies of energy materials. *Philosophical Transactions of the Royal Society A* 368, no.1923 (July): 3379–3456. doi:10.1098/rsta.2010.0111.

Chen, X., S. Shen, L. Guo, and S. M. Mao. 2010. Semiconductor-based photocatalytic hydrogen generation. *Chemical Reviews* 110, no.11 (November): 6503. doi:10.1021/cr1001645.

Damaskin, B. B., and O. A. Petrii. 2011. Historical development of theories of the electrochemical double layer. *Journal of Solid State Electro-Chemistry* 15: 1317–1334. doi:10.1007/s10008-011-1294-y.

de Angelis, F., S. Fantacci, and A. Selloni. 2008. Alignment of the dye's molecular levels with the TiO_2 band edges in dye-sensitized solar cells: a DFT–TDDFT study. *Nanotechnology* 19, no.42 (October): 424002. doi:10.1088/0957-4484/19/42/424002.

Hwang, D. W., J. S. Lee, W. Li, and S. H. Oh. 2003. Electronic band structure and photocatalytic activity of $Ln_2Ti_2O_7$(Ln=La, Pr, Nd). *Journal of Physical Chemistry B* 107, no.21 (May): 4963–4970. doi:10.1021/jp034229n.

Jeyalakshmi, V., S. Tamilmani, R. Mahalakshmy, P. Bhyrappa, K. R. Krishanmurthy, and B. Viswanathan. 2016. Sensitization of La modified $NaTaO_3$ with cobalt tetra phenyl porphyrin for photo-catalytic reduction of CO_2 by water with UV-visible light. *Journal Molecular Catalysis A*, 420 (August): 200–207. doi:10.1016/j.molcata.2016.04.027 and references therein.

Kalyanasundaram, K., and M. Graetzel. 2010. Artificial photosynthesis: biomimetic approaches to solar energy conversion and storage. *Current Opinion on Biotechnology* 21, no.3 (June): 298–310. doi:10.1016/j.copbio.2010.03.021 and references therein.

Kato, H., K. Asakura, and A. Kudo. 2003. Highly efficient water splitting into H_2 and O_2 over lanthanum-doped $NaTaO_3$ photo-catalysts with high crystallinity and surface nanostructure. *Journal of American Chemical Society* 125, no.10 (February): 3082–3089. doi:10.1021/ja027751g.

Kullgren, J., B. Aradi, T. Frauenheim, L. Kavan and P. Deak. 2015. Resolving the controversy about the band alignment between Rutile and anatase: The role of OH-/H+ adsorption. *Journal of Physical Chemistry* C 119:21952–21958.

Mi, Y., and Y. Weng. 2015. Band alignment and controllable electron migration between rutile and anatase TiO_2. *Scientific Reports* 5, 11482 (July). doi:10.1038/srep11482.

Minggu, L. J., W. R. W. Daud, and M. B. Kassim. 2010. An overview of photocells and photoreactors for photoelectrochemical water splitting. *International Journal of Hydrogen Energy* 35, no.11 (June): 5233–5244. doi:10.1016/j.ijhydene.2010.02.133.

Osterloh, F. E. 2008. Inorganic materials as catalysts for photochemical splitting of water. *Chemistry of Materials* 20, no. 1 (January): 35–54. doi:10.1021/cm7024203.

Osterloh, F. E. 2013. Inorganic nanostructures for photo-electrochemical and photocatalytic water splitting. *Chemical Society Review* 42 (October): 2294–2320. doi: 10.1039/C2CS35266D.

Osterloh, F. E., and B. A. Parkinson. 2011. Recent developments in solar water-splitting photocatalysis. *Materials Research S Bulletin* 36, no.1: 17–22. doi:http://dx.doi.org/10.1557/mrs.2010.5.

Peng, R., C. M. Wu, J. Baltrusaitis, N. M. Dimitrijevic, T. Rajh, and R. T. Koodali. 2013. Ultra-stable CdS incorporated Ti-MCM-48 mesoporous materials for efficient photocatalytic decomposition of water under visible light illumination. *Chemical Communications* 49, no. 31 (April): 3221–3223. doi:10.1039/C3CC41362D.

Saito, K., K. Kogaa, and A. Kudo. 2011. Lithium niobate nanowires for photocatalytic water splitting. *Dalton Transaction* 40, no.15 (April): 3909–3913. doi:10.1039/C0DT01844A.

Sato, J., N. Saito, H. Nishiyama, and Y. Inoue. 2003. Photocatalytic activity for water decomposition of indates with octahedrally coordinated d^{10} configuration. Influences of preparation conditions on activity. *Journal of Physical Chemistry B* 107, no.31 (July): 7965–7969. doi:10.1021/jp030020y.

Scanlon, D. O., C. W. Dunnill, J. Buckeridge, et al. 2013. Band alignment of rutile and anatase TiO_2. *Nature Materials* 12 (July): 798–801. doi:10.1038/nmat3697.

Takata, T., and K. Domen. 2009. Defect engineering of photocatalysts by doping of aliovalent metal cations for efficient water splitting. *Journal of Physical Chemistry C* 113:19386–19388. doi:10.1021/jp908621e.

Townsend, T. K., N. D. Browning, and F. E. Osterloh. 2012. Nanoscale strontium titanate photocatalysts for overall water splitting. *ACS Nano* 6, no. 8 (July): 7420–7426. doi:10.1021/nn302647u.

Viswanathan, B. 2003. Photocatalytic processes—selection criteria for the choice of materials. *Bulletin of the Catalysis Society of India* 2: 71. https://nccr.iitm.ac.in/251_B.Viswanathan.pdf.

Viswanathan, B. 2005. Photo/electrochemistry & photobiology in the environment, energy and fuel. In *Research Sign Post*, eds. S. Kaneco, B. Viswanathan, and Kunihiro Funasaka: 1–12, Research Sign Post, Trivandrum, India.

Viswanathan, B., and A. Scibioh. 2009. *Fuel Cells Principles and Applications*. Hyderabad, India: University Press.

Walter, M. G., E. L. Warren, J. R. Mckone, et al. 2010. Solar water splitting cells. *Chemical Reviews* 110, no.11 (November): 6446–6473. doi:10.1021/cr1002326.

Yan, S., L. Wan, Z. Li, and Z. Zou. 2011. Facile temperature-controlled synthesis of hexagonal Zn_2GeO_4 nanorods with different aspect ratios toward improved photocatalytic activity for overall water splitting and photoreduction of CO_2. *Chemical Communications* 47 (April): 5632–5634. doi:10.1039/C1CC10513B.

4

Heterostructured Photocatalysts for Solar Fuel Generation

Kamala Kanta Nanda and Yatendra S. Chaudhary

CONTENTS

4.1 Introduction

Photocatalysis consists of various intriguing processes: excitation, separation, and transport of photogenerated charge carriers (excitons) to the photocatalyst–electrolyte interface to drive the fuel-forming reactions as illustrated in Figure 4.1. The details of processes taking place at the molecular level and the photochemistry involved in the photocatalytic water splitting are discussed in detail in Sections 2.1, 2.3, and 3.1.

Ever since the pioneering demonstration of the water splitting using the TiO_2 semiconductor photocatalyst by Fujishima and Honda (1972), a variety of strategies are being explored to tune the fundamental processes. However, there has been incremental progress in the overall photocatalytic water splitting efficiency. One of the highest solar-to-hydrogen (STH) conversion efficiency showed by a single band gap semiconductor photocatalyst ($BiVO_4$) is about 1.8% (Kuang et al. 2016). In order to further the efficiency, a multi-bandgap based tandem strategy is being adapted. A tandem cell consisting

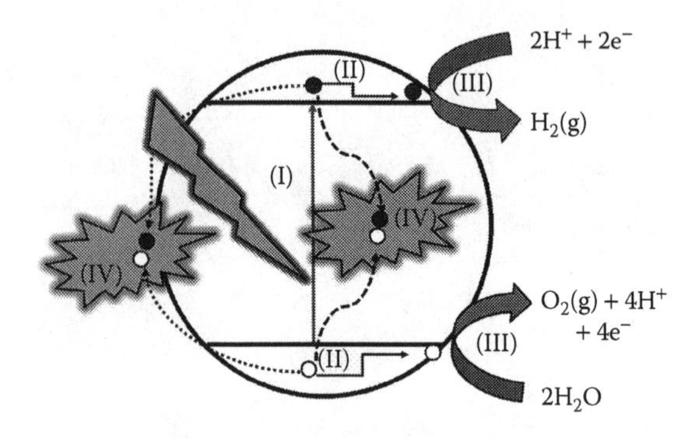

FIGURE 4.1
Photocatalytic water splitting with sunlight consists of three main steps: (I) a semiconductor absorbs light photons and generates excited electrons and holes; (II) these excited electrons and holes can migrate to the surface of the semiconductor; (III) on the surface, holes can oxidize water to O_2 and electrons can reduce protons to H_2; (IV) electrons and holes may recombine inside the bulk and the surface of the photocatalyst.

of GaInP n–p top and a GaInAs n–i–p bottom cell showed the STH of about 14%, the highest efficiency reported till date (May et al. 2015).

One of the major cause that limits the overall efficiency is fast recombination of excitons, as the recombination process (~10^{-9} s) is much faster than that of the electron diffusion to the photocatalyst surface reaction site (10^{-8} to 10^{-1} s) (Yuan et al. 2014). Further, it is difficult for a single band gap photocatalyst to possess both wide light absorption range and strong redox ability. To overcome these limitations, the strategy to form heterostructured photocatalysts is desirable. In such heterostructured photocatalysts, the potential gradient is generated at the heterointerface due to the offset of band edges of both the semiconductors. This potential gradient generated across the heterointerface results in better charge separation. Such a configuration of heterostructured photocatalysts mimics the intermediate pathways (Z-scheme) of the biological photosynthesis. Thereby, a tremendous interest has grown to develop heterostructured photocatalysts in recent years.

4.2 Heterojunction Architecture

Appropriate band edge alignment with right energetic of two semiconductors is important to design an efficient heterostructured photocatalyst (Figure 4.2). There are three possible ways to align band edges of two semiconductors. Type I: the conduction band edge (E_c) of semiconductor B is higher than that of semiconductor A. The valance band edge (E_v) of semiconductor B is lower than that of semiconductor A; therefore, holes

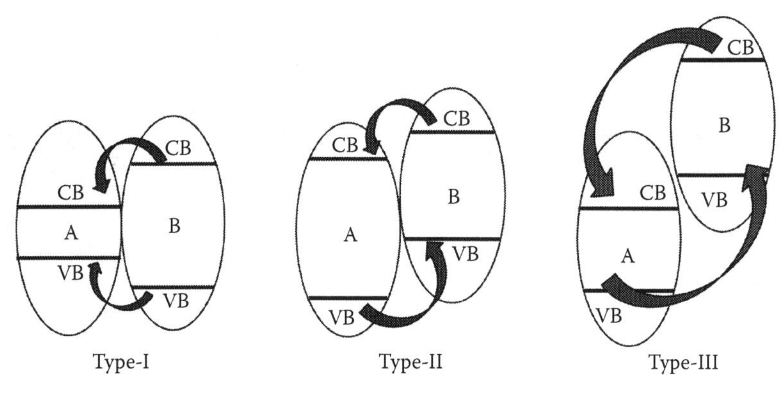

FIGURE 4.2
Band alignment in types I, II, and III heterojunctions.

and electrons are transferred and accumulated in semiconductor A. Type II: the electrons are transferred from semiconductor B to A due to more negative E_c of semiconductor B, whereas holes are transferred in opposite direction from the more positive E_v of semiconductor A to B (Marschall 2014). Such an alignment enables efficient charge separation across the heterointerface. Type III: the band gap of two semiconductors does not overlap and the exciton transfer is similar to type II except the fact that there is larger difference in E_v and E_c positions of semiconductors A and B, resulting in higher driving force for charge transfer.

4.3 State of the Art: Charge Carrier Separation at Heterointerface

The migration of excitons in opposite direction in type II spatially isolates them, allowing faster separation of photogenerated electrons and holes (Robel et al. 2007). On the other hand, the redox ability of photogenerated electrons and holes is weakened as the E_v of B is less positive than that of A and the E_c of A is less negative than that of B (Figure 4.3). Such a band edge alignment thus cannot concurrently possess improved charge separation efficiency and strong redox ability. Hence, there is a need to explore a new-type heterostructure to address the aforementioned issues. Inspired from the Z-scheme mechanism of biological photosynthesis (as discussed in Section 1.2), artificial Z-scheme photocatalytic systems are being designed that possess strong redoxability of the electrons and/or holes in the respective components. In artificial Z-scheme systems, a redox couple (electron acceptor/donor, A/D) is generally used as the redox mediator. However, such A/D couple also favors the backward reaction, decreasing the effective

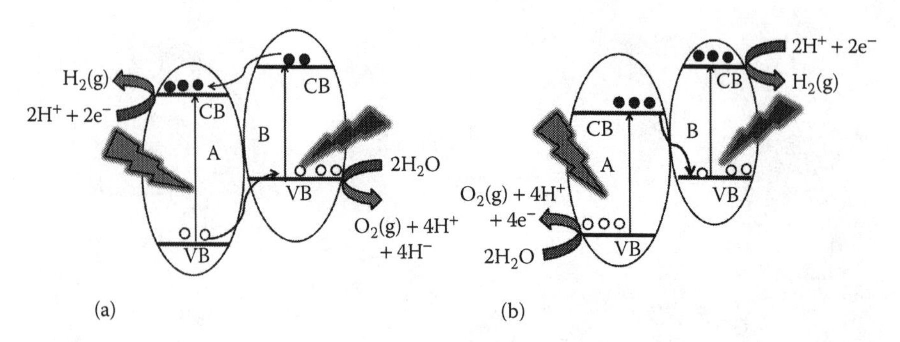

FIGURE 4.3
Schematic showing the charge carrier separation in (a) type-II heterojunction (b) Z-scheme of photocatalyst.

number of photogenerated electrons and holes, thereby lowering the efficiency (Sayama et al. 2002).

In order to avoid this, the solid-state Z-scheme can be designed by forming a suitable heterointerface between two semiconductors having right-band edge energetics or a suitable conductor as electron mediator between two semiconductor photocatalysts. In the Z-scheme heterostructured photocatalytic system, simultaneous photoexcitation of electrons in two components takes place (Figure 4.3b). The excited electron from conduction band (CB) of semiconductor A (which is less negative than CB of semiconductor B) is transferred to the valence band (VB) semiconductor B (which is less negative than VB of semiconductor A) via a sacrificial recombination, leading to the spatially separated electrons and holes. Unlike type II heterojunction, the Z-scheme allows the spatial separation of excitons without trading off with their redox potentials.

The number of electrons in CB of semiconductor A and number of holes in VB of semiconductor B must be equal so that the effective sacrificial recombination between them occurs and an effective electron–hole separation is achieved. When the photogenerated charge carrier in one of the semiconductor is higher than that of the other, the surplus number of these charge carriers may recombine in the bulk of respective semiconductors. The ideal situation for the same number of charge carrier generation by two semiconductors can be achieved by optimizing the molar ratio of semiconductors A and B that is also affected by the geometry. For example, the Z-scheme $BiVO_4$-Ru/$SrTiO_3$: Rh system was synthesized by mixing $BiVO_4$ and Ru/$SrTiO_3$: Rh powders in the aqueous solution. The highest photocatalytic activity was achieved when the effective Z-scheme heterostructure formed at the mass ratio between $BiVO_4$:Ru/$SrTiO_3$: Rh of 1:1 (Sasaki et al. 2009). Whereas in the TiO_2-C_3N_4 Z-scheme system, the highest photodegradation of formaldehyde was achieved when the mass ratio between C_3N_4 and TiO_2 was 3:22 (Yu et al. 2013a). Further, if the mass ratio is lower than that of 3:22, the heterointerface area decreases, which has detrimental effect on charge carrier separation

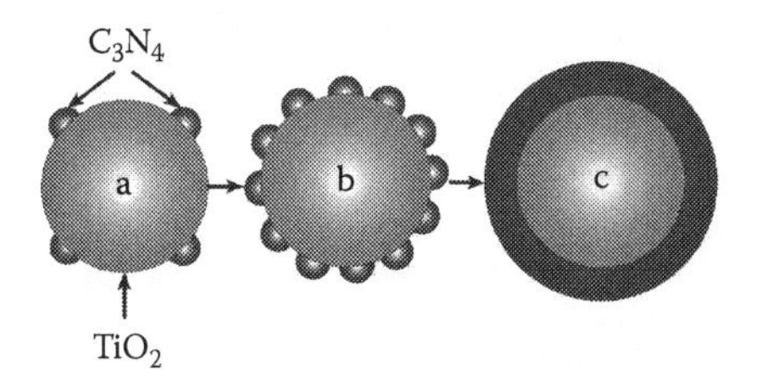

FIGURE 4.4
Architectures of the TiO_2-C_3N_4 systems with different mass ratios. (Zhou, P. et al.: All-solid-state Z-scheme photocatalytic systems. *Advanced Materials*, 2014, 26, 4920–4935. Copyright Wiley-VCH Verlag GmbH & Co. KGaA. Reproduced with permission.)

and thus lowers degradation efficiency. If the mass ratio is higher than that of 3:22, there will be a formation of core–shell TiO_2-C_3N_4 structure (Figure 4.4). The thicker C_3N_4 layer will shield TiO_2 from harvesting solar radiation, consequently leading to the generation of unequal number of charge carriers. This will also extend the electron transfer distance from TiO_2 CB to C_3N_4 VB, which further increases the probability of the recombination of excitons.

It is difficult to envisage the mechanism of solar water splitting in the heterostructured photocatalyst due to the similarity in the heterointerface of type II heterojunction and Z-scheme heterostructures. Nonetheless, the Z-scheme electron transfer can be confirmed, if the following conditions are met by the heterostructured photocatalysts.

1. Semiconductor A can only generate O_2 in the presence of electron acceptor.
2. Semiconductor B can only produce H_2 in the presence of electron donor.
3. The overall water splitting can occur simultaneously in the absence of electron acceptor and donor.

The unique electron transfer pathway followed by Z-scheme photocatalytic systems has tempted to widely explore a variety of combinations to form heterostructured photocatalysts (Zhou et al. 2014).

4.4 Heterojunction Photocatalysts

Various kinds of combinations of semiconductor photocatalysts are being explored to design heterojunction photocatalysts that are discussed under three categories on the basis of the nature of materials: (1) inorganic–inorganic (referred as I-I here onward), (2) inorganic–organic (referred as I-O here onward), (3) organic–organic (referred as O-O here onward) heterostructured photocatalysts. The designing aspects of heterostructured photocatalysts and their influence on the photocatalytic activity are elaborated in the following section.

4.4.1 Inorganic–Inorganic Heterostructured Photocatalysts

To overcome fast charge carrier recombination, photocorrosion in narrow band gap semiconductors, and limited visible light absorption by wide band gap semiconductors, a variety of I-I heterostructured photocatalysts have been developed. These consist of either p–n or n–n heterojunction.

TiO_2 has been widely used for photoelectrochemical water splitting and still the most commonly used photocatalyst with modifications. Owing to its wide band gap, it can only harvest the ultraviolet (UV) light. This leads to a very low theoretical maximum solar-to-H_2 conversion (STH) efficiency (η = 1.3% for anatase and 2.2% for rutile TiO_2) (Li and Wu 2015).

The modification of TiO_2 with $BiVO_4$ to form heterostructured photocatalysts showed long lifetime (~3 ms) of photoexcited charge carriers and hence improved charge separation (Xie et al. 2014). It is mainly due to faster transfer of electrons from $BiVO_4$ to TiO_2. The optimized $TiO_2/BiVO_4$ with a molar ratio of 5% produced 22 μmol h^{-1} g^{-1} when 20 vol% methanol–water mixture was used (Ng et al. 2010).

Although TiO_2 and CdS are widely studied photocatalysts, they suffer from either inability to harvest light in visible region, shorter lifetime of excitons, or photocorrosion, respectively. The combination of both to form heterostructured photocatalysts may subdue these hitches for photocatalytic H_2 production. In CdS-TiO_2 heterostructured photocatalysts, excited electron moves from CdS to TiO_2 owing to the more electronegative E_c. This leads to better charge separation and thus higher quantum yield. The holes in the CdS are quenched by sacrificial agents, curtailing the photocorrosion of CdS. Ultra-thin TiO_2 nanodiscs were modified by CdS nanoparticles as visible light sensitizer and Ni nanoparticles as cocatalyst. This heterostructure produced H_2 under visible light irradiation with apparent quantum yield (AQY) of 22.4% at λ = 420 nm at the rate 15,326 μmol h^{-1} g^{-1} (Dinh et al. 2014).

There are some reports on ternary heterostructures employed for solar fuel generation (Fang et al. 2013; Xinga et al. 2015; Yu et al. 2015). Au@TiO_2-CdS ternary nanostructure was synthesized by grafting CdS nanoparticles onto Au@TiO_2 core–shell structures (Fang et al. 2013). The Au@TiO_2-CdS exhibit 10 times higher solar H_2 production as compared to that of binary TiO_2/CdS. The authors attributed the improved photocatalytic activity to the unique transfer pathway (CdS → TiO_2 → Au) that builds up in these ternary heterostructures; photoexcited electrons of CdS are transferred to the core of Au nanoparticles via TiO_2 nanocrystal bridge. This effectively diminishes the electron–hole recombination in the CdS photocatalyst.

ZnO/CdS heterostructure photocatalyst follows the charge carrier transfer mechanism similar to the Z-scheme, as discussed in Section 4.3, where the recombination of the photoexcited electron from the ZnO conduction band and hole from the CdS valence band occurs at the interface (Wang et al. 2009, 2012; Yu et al. 2013b). In this case, the photoexcited electrons in CdS with more negative E_c and holes in ZnO with more positive E_v drive the water splitting to produce H_2 with an enhanced rate. To design ternary heterostructure CdS/Au/ZnO, a core–shell structure of Au/CdS was selectively deposited on the polar surface of ZnO, through a two-step self-assembly process (Yu et al. 2013b). To fabricate the Z-scheme ZnO-Au-CdS system, the precursor (Au^{3+}) selectively reduced and deposited on the electron-rich Zn^{2+}-terminated {001} surface of hexagonal ZnO nanosheets under UV irradiation (Figure 4.5a and b; Yu et al. 2013b). CdS shells grew around Au cores owing to the affinity of S^{2-} ions to the Au cores, by a chemical deposition method. The nanojunctions formed in ZnO-Au-CdS are shown in Figure 4.5c and d. CdS/Au/ZnO heterostructured photocatalyst showed remarkable enhancement in the photocatalytic H_2 evolution by a factor of 4.5 (608 μmol h^{-1} g^{-1}) than that of the binary CdS/ZnO (134 μmol h^{-1} g^{-1}) heterostructured photocatalyst (Figure 4.5f).

CdS-$ZnWO_4$ forms a type I heterojunction, which generates 31.46 mmol h^{-1} g^{-1} of H_2 under visible light illumination that is approximately 8 and 755 times higher than that obtained with bare CdS and bare $ZnWO_4$ under similar conditions (Xu et al. 2015). Both $ZnWO_4$ and CdS are n-type semiconductors. The E_c of $ZnWO_4$ is more negative (−0.65 V vs NHE at pH 7) than that of CdS

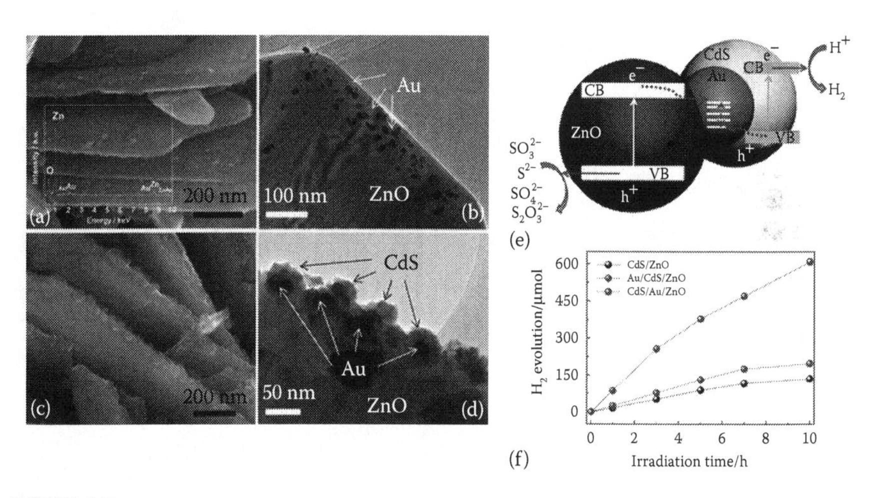

FIGURE 4.5
SEM and TEM images of Au-ZnO (a and b); SEM and TEM images of the ZnO-Au-CdS heterostructures (c and d). The inset in (a) is the EDS spectrum of Au-ZnO; the vectorial Z-scheme charge carrier transfer process in the CdS/Au/ZnO heterostructure (e) and time courses of photocatalytic H_2 evolution from water with the CdS/ZnO, Au/CdS/ZnO, and CdS/Au/ZnO photocatalysts. Reaction conditions: 100 mg of photocatalyst was suspended in 270 mL aqueous solution with 0.1 M Na_2SO_3 and 0.1 M Na_2S (f). (Yu, Z. B. et al., *Journal of Materials Chemistry A*, 1, 2773–2776, 2013b. Reproduced by permission of the Royal Society of Chemistry.)

(−0.4 V vs NHE at pH 7) and the E_v of $ZnWO_4$ (2.57 V vs NHE at pH 7) is more positive than that of CdS (1.43 V vs NHE at pH 7). When a n–n CdS-$ZnWO_4$ heterojunction is formed, the electrons move from $ZnWO_4$ to CdS because of more negative E_c of $ZnWO_4$. This results in the formation of depletion layer (positive region) on $ZnWO_4$ and an accumulation layer (negatively charged region) on CdS (Figure 4.6b). When the process reaches equilibrium state (equalized Fermi level, E_f), a built-in electric field is established at the heterojunction. On irradiation with visible light, the photogenerated charge carriers

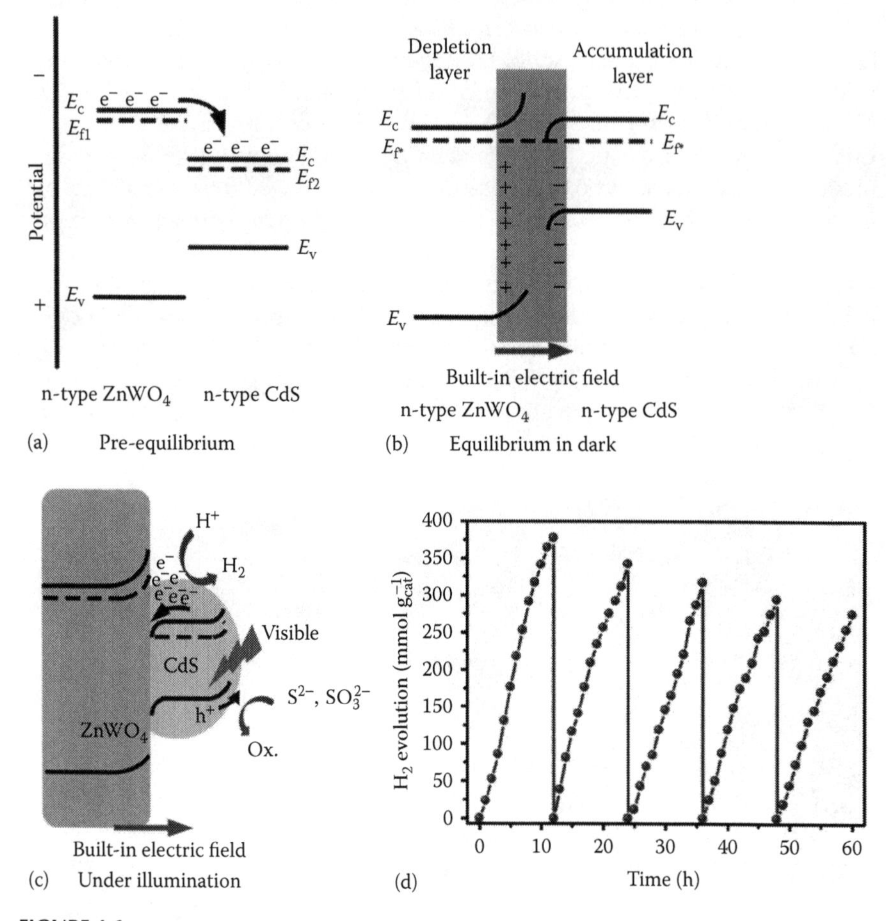

FIGURE 4.6
Schematic diagram of the band configuration (a) and the charge separation at the heterointerface of CdS-$ZnWO_4$ heterojunctions under visible light irradiation (b and c); (d) time course of hydrogen evolution over 20 mg of nonloaded CdS-$ZnWO_4$ heterojunction under visible light irradiation ($\lambda > 420$ nm) for 60 h. E_c is the lowest energy level of conduction band, E_v the highest energy level of valance band, E_{f1} and E_{f2} are the Fermi levels of $ZnWO_4$ and CdS, E_f is the Fermi level of CdS-$ZnWO_4$ heterojunctions. (Xu, M. et al.: Rationally designed n–n heterojunction with highly efficient solar hydrogen evolution, *ChemSusChem*, 2015, 8, 1218–1225. Copyright Wiley-VCH Verlag GmbH & Co. KGaA. Reproduced with permission.)

are generated in CdS. Their separation is assisted by the built-in electric field, promoting electron transfer toward $ZnWO_4$. However, the electrons cannot migrate to $ZnWO_4$ because of more negative E_c of $ZnWO_4$, thereby leading to the accumulation of photogenerated electrons at the interface of $ZnWO_4$ and CdS (Figure 4.6c). These electrons reduce H^+ to H_2. At the same time, the photogenerated holes in CdS migrate in the opposite direction and react with sacrificial agents. These phenomena occurring in unison also suppress the photocorrosion of CdS as revealed by the fact that 73% activity retained even after the 60 h continuous photocatalytic H_2 generation (Figure 4.6d).

Niobate photocatalysts are more promising than titanates due to their rich polymorphism over a wide range of temperatures, high electrooptic coefficients, optical damage resistance, and more reductive conduction band potential than most of the oxides. However, the wide band gap (3.4 eV—allows to harvest only UV radiation) curtails their significance for practical solar H_2 generation application. The $NaNbO_3$ was modified with CdS such that it can harvest visible solar radiation and exploit the heterointerface for improving charge separation. While coupling CdS with $NaNbO_3$, the influence of the morphology of photocatalyst particles over the charge separation phenomenon was also considered (these aspects have been elaborated in Section 3.1). Keeping in mind that the vectorial transfer of photoexcited electron in nanorods reduces recombination probability of excitons, heterostructured CdS-$NaNbO_3$ nanorods were synthesized. The photocurrent density (current density under light – current density under dark) measurements undertaken reveal that the CdS/$NaNbO_3$ nanorod heterostructured photocatalyst exhibits drastically enhanced photocurrent (J_{an}: 0.85 mA cm^{-2} at 0.6 V vs SHE) as compared to that of bare CdS (J_{an}: 0.27 mA cm^{-2} at 0.6 V vs SHE) and bare

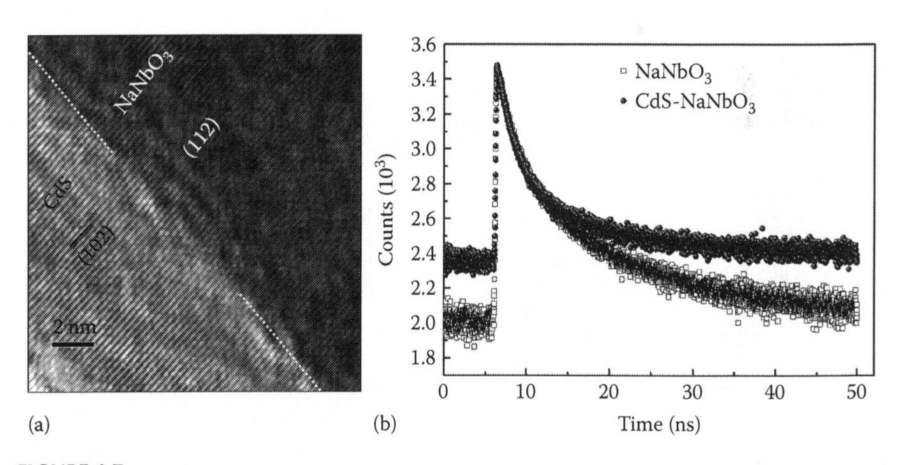

FIGURE 4.7
(a) HRTEM image showing the lattice matched heterointerface in CdS/$NaNbO_3$ nanorod, (b) time-resolved photoluminescence spectra of CdS/$NaNbO_3$ nanorod photocatalyst, and $NaNbO_3$ nanorod photocatalyst. (Nanda, K. K. et al., *RSC Advances*, 4, 10928–10934, 2014. Reproduced by permission of the Royal Society of Chemistry.)

$NaNbO_3$ nanorod (J_{an}: 0.17 mA cm^{-2} at 0.6 V vs SHE). The detailed High resolution transmission electron microscopy (HRTEM) and time-resolved photoluminescence decay measurements show the formation of lattice matched heterointerface (which is not so common to achieve) and longer average lifetime of excitons (8.06 ns) in the case of CdS/$NaNbO_3$. On the other hand, the average lifetime of excitons for unmodified $NaNbO_3$ nanorod is 6.45 ns (Figure 4.7). From these results, it was inferred that the unison of lattice-matched heterointerface, longer lifetime of excitons, and nanorod morphology leading to improved charge separation across the CdS/$NaNbO_3$ heterointerface and are responsible for such improvement in the photocatalytic activity (Nanda et al. 2014).

4.4.2 Inorganic–Organic Heterostructured Photocatalysts

In order to exploit the synergized properties (such as high absorption coefficient in visible light, high charge carrier mobility, and high dielectric permittivity) arising from both the organic and inorganic counterparts, the organic–inorganic heterostructured photocatalysts are being designed.

The poly(3-hexylthiophene) (P3HT) that is a hole transporting polymer was coupled with CdS for designing organic–inorganic heterostructured photocatalysts. Under optimized conditions, the P3HT/CdS heterostructured photocatalyst shows enhanced solar H_2 generation than that shown by unmodified CdS (Figure 4.8). Such an improvement in the solar H_2 generation activity has been attributed to the unison of shorter lifetime of excitons (due to the efficient charge separation across the heterointerface), higher degree of band bending, and increased donor density (Nanda et al. 2015).

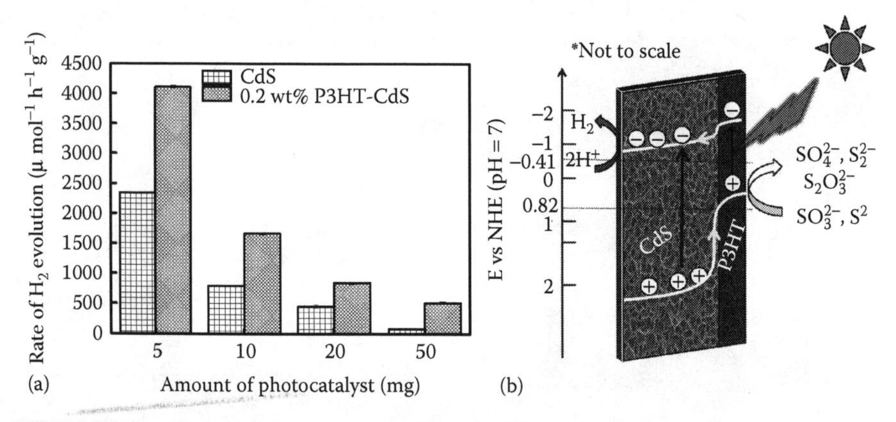

FIGURE 4.8
(a) Photocatalytic H_2 evolution rate for unmodified CdS and 0.2 wt% P3HT/CdS (8 h, 5 mL 0.1 M Na_2S–0.14 M Na_2SO_3, pH ~ 12.4). (b) Schematic representation of the photocatalytic hydrogen generation on heterostructured P3HT/CdS photocatalyst. (Reprinted with permission from Nanda, K. K. et al., *ACS Applied Materials & Interfaces*, 7, 7970–7978, 2015. Copyright [2015] American Chemical Society.)

The isotype heterojunctions of graphitic carbon nitride (CN) and sulfur-mediated graphitic carbon nitride (CNS) were designed with close interconnection between CN and CNS by surface-assisted polymerization (Zhang et al. 2012). One was CNS-CN (CN serving as the host), and the other was CN-CNS (CNS serving as the host). The optimized CNS-CN isotype heterostructure exhibited ~11 and ~2.3 times higher solar H_2 production than that observed with CN and CNS, respectively. The photogenerated electrons are transferred from CN (E_c: −1.42 V) to CNS (E_c: −1.21 V), driven by the CB offset, and the VB offset between CNS (E_v: +1.46 V) and CN (E_v: +1.28 V) could induce the migration of photoinduced holes from CNS to CN, giving rise to the redistribution of electrons on the CNS side and holes on the CN side at the heterointerface (Figure 4.9), which in turn greatly reduces the energy-wasteful exciton recombination. In addition, the lifetime of excitons was also prolonged due to the staggered band offsets, increasing the probability of electrons or holes being available to drive redox reactions.

4.4.3 Organic–Organic Heterostructured Photocatalysts

In order to exploit the hole-transporting property of P3HT, it has been coupled with graphitic carbon nitride (gC_3N_4) to form a heterojunction. The 3 wt% P3HT/gC_3N_4 heterostructured photocatalyst exhibited high apparent quantum yield (AQY) of 77.4% at 420 nm (Zhang et al. 2015). It varies from 59.4%, 20.2%, 3.2%, to 0.68% at 500, 600, 700, and 800 nm monochromatic light irradiation, respectively. When there was no sacrificial agent (ascorbic acid in this case), the H_2 generation rate decreased from 304.5 to 0.625 mmol $h^{-1}g^{-1}$. The improved photocatalytic H_2 generation is due to the efficient exciton transfer across the heterointerface of P3HT/g-C_3N_4, as revealed by the longer lifetime of excitons (2.02 ns) than that observed for bare P3HT (1.11 ns).

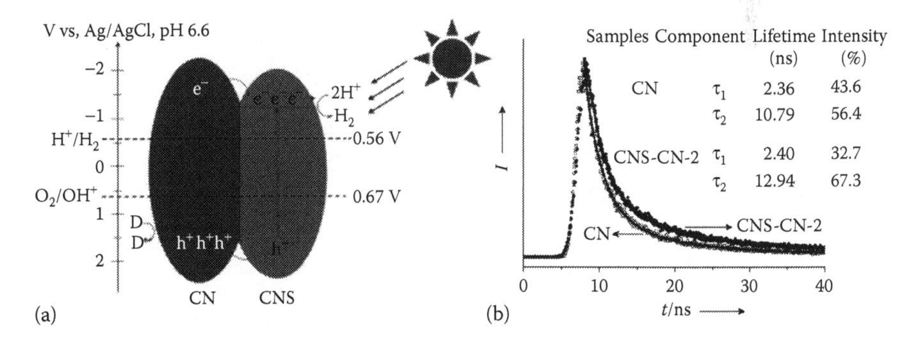

FIGURE 4.9
(a) Schematic illustration of organic heterojunction formed between CN and CNS. D = donor; (b) time-resolved PL spectra monitored at 480 nm under 420 nm excitation at 77 K for CN and CNS–CN. (Zhang, J. et al.: A facile band alignment of polymeric carbon nitride semiconductors to construct isotype heterojunctions, *Angewandte Chemie International Edition*, 2012, 51, 10145–10149. Copyright Wiley-VCH Verlag GmbH & Co. KGaA. Reproduced with permission.)

There are handful reports in the literature on O-I and O-O heterostructured photocatalysts that form heterojunctions. It is possibly due to the sensitivity of the organic component to oxidation, which compromises an important stability aspect of the photocatalysts.

Very recently, the computational studies have shown the formation of van der Waals' heterojunction using metal-free 2D sustainable materials (boron nitrides, carbon nitrides) in combination with ZrS_2 nanosheets. It relies on the fact that by applying slight strain, E_c and E_v can be tuned to favorably alter the charge separation and driving force to drive fuel-forming reaction (Zhang et al. 2016). However, this concept is yet to be tested.

4.5 Molecular Hybrid Photocatalysts

Inspired by the naturally occurring enzymes, their high catalytic efficiency (turnover frequency 10^3–10^4 s^{-1}) with minimal overpotential requirements, a class of catalysts commonly known as "molecular catalyst" has recently gained attention (Svetlitchnyi et al. 2001). In order to drive H_2 evolution reaction using solar radiation, they are coupled with light-harvesting component—semiconductor or dye. Although the molecular catalyst–semiconductor heterostructures do not form the heterojunction, as discussed in this chapter earlier, a brief overview of these molecular hybrid photocatalysts is presented below.

One of the early examples on solar hydrogen production using a molecular sensitizer was reported in 1981 by Borgarello et al. This bifunctional redox catalyst consists of Pt, RuO_2 codeposited on colloidal TiO_2, sensitizer Ru $(bipy)_3^{2+}$, and an electron relay molecule methyl viologen. Here, the excitation of sensitizer is followed by the electron injection into CB of TiO_2. Then the electron is readily transferred to the active site (Pt on TiO_2) to reduce H^+ to H_2 at the rate of 0.34 mL H_2 g^{-1}. The sensitizer is regenerated in situ and is coupled to oxygen-evolving reaction at the RuO_2 catalytic site.

To eliminate the use of an expensive noble metal cocatalyst Pt, a molecular cobalt H_2 evolution catalyst (Et_3NH) [$Co^{III}(dmgH)_2$(pyridyl-4-hydrophosphonate)] (Et_3NH) [CoP1] was used in the visible light-driven H_2 production photocatalytic system containing a Ru (II) dye (ruthenium tris-bipyridine-based dye—RuP) as a sensitizer, triethanolamine as an electron donor, and both COP1 and RuP were immobilized on TiO_2 via the phosphonic acid linkage (Lakadamyali et al. 2012). TiO_2 acts as an electron transfer mediator that controls the generation and transfer of electrons from RuP to the molecular catalyst and hence plays an important role in facilitating the separation of electrons and holes. On excitation with visible light, RuP gets excited and injects its excited electron into the CB of TiO_2 at a time scale of 180 ps. Subsequently, RuP is regenerated by the oxidation of triethanolamine and the process occurs in a time scale of 0.1 ms. In the next step, the electron that was transferred from RuP to TiO_2 in turn injected into the molecular catalyst CoP1 in 10 μs, which is about 100 times faster than that of the charge carrier recombination with the oxidized RuP, thereby reducing the oxidation state of Co from +3 to +1 by two one-electron reduction steps that subsequently produces H_2, as shown in Figure 4.10. Furthermore, this process is thermodynamically favorable as evident from the overpotential data. The overpotential of 0.3 V ($E^*_{RuP} = -0.95$ V, E_c [TiO_2] = 0.7 V vs NHE at pH 7) on TiO_2 available for the reduction of H^+ is sufficient for CoP1, which reduces H^+ to H_2 at an overpotential of 0.2 V. These favorable processes resulted in the turnover number (TON) of about 108 (±9) mol H_2 (mol CoP1)$^{-1}$ that corresponds to a quantum efficiency of about 1%.

Another analogous hybrid photocatalytic system, consisting of g-C_3N_4 as a sensitizer, end group functionalized cobaloxime complexes as molecular catalyst (carboxy-functionalized cobaloxime, C1; pyrene-functionalized cobaloxime, C2; and nonfunctionalized cobaloxime, C3), and triethanolamine as electron donor (Song et al. 2014). Here the authors pointed out that the highest TON was obtained when the volume ratio of triethanolamine:water is 9:1. It is due to the fact that the protonation of triethanolamine takes place if there is high water content that drastically decreases its reduction power, which in turn affects the regeneration of the sensitizer (Ishida et al. 1990). The higher TON (281) for H_2 generation recorded for the C2/g-C_3N_4 system is due to stronger π–π interactions between the pyrene moiety and g-C_3N_4, which facilitate faster transfer of the electron to the molecular catalyst.

4.6 Biomimetic Photocatalysts

Enzymes are the best catalysts available in nature. Hydrogenase, a class of metalloenzymes, catalyzes the reversible conversion of H^+ to H_2 in biological systems (Vignais and Billoud 2007). The active sites of these enzymes contain nonprecious metals (nickel and/or iron) that are buried deeply within the protein, and chains of Fe-S clusters mediate long-range electron transfer to/or from a redox partner. Coupling the catalytic power of enzymes with semiconductor nanomaterials can effectively reduce H^+ to H_2. In this system, semiconductor can act both as the light harvester and electron mediator. The surface functionalization of the semiconductor can be done using various ligands to couple with enzyme in an electroactive manner. The passivation effect of the ligand can also intensely affect the photophysical properties of the semiconductor and energetics of external charge transfer. The length, degree of bond saturation (aromaticity), and solvent-exposed functional groups of ligands can be manipulated to further tune the interface to control molecular assembly and stability of semiconductor–enzyme hybrid.

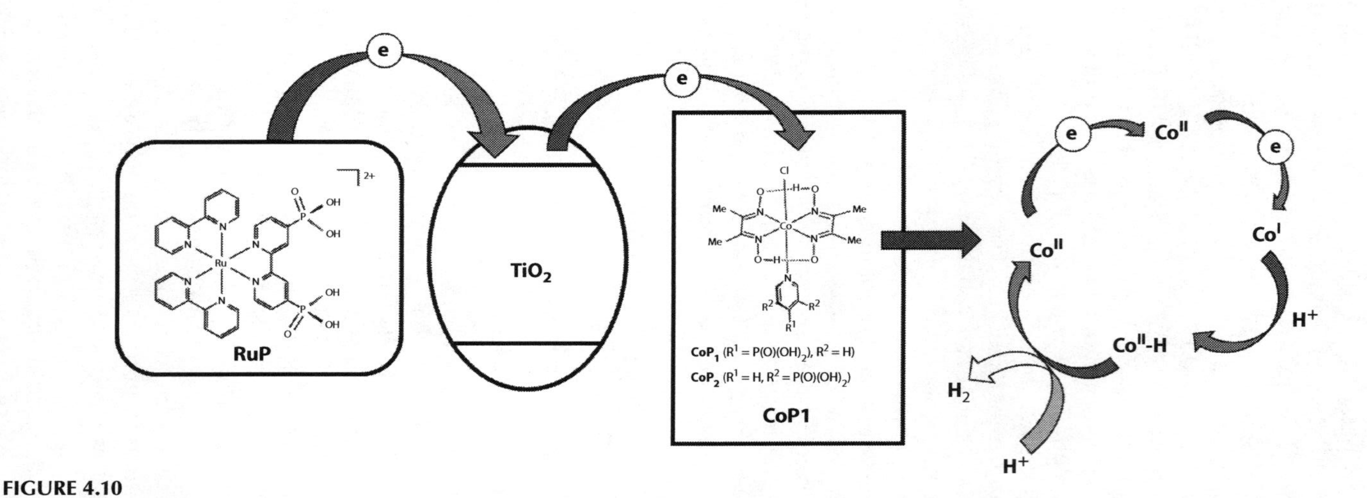

FIGURE 4.10
Schematic representation of photocatalytic H_2 evolution with RuP/CoP1 modified TiO_2. (Lakadamyali, F. et al., Electron transfer in dye-sensitised semiconductors modified with molecular Cobalt Catalysts: photoreduction of aqueous protons, *Chemistry-An European Journal*, 2012, 18, 15464–15475. Copyright Wiley-VCH Verlag GmbH & Co. KGaA. Reproduced with permission.)

One of the first examples of solar H_2 production by a hydrogenase-semiconducting material hybrid was with [FeFe]-hydrogenase and TiO_2 (Cuendet et al. 1986). In this study, three different hydrogenase enzymes (isolated from *Clostridium pasteurianum, Desulfovibrio desulfuricans* strain Norway 4, and *D. baculatus* 9974) were introduced into the suspension of anatase TiO_2. These enzyme–semiconductor hybrid systems were able to produce H_2 under band gap illumination of the semiconducting particles in the presence of electron donors such as Ethylenediaminetetraacetic acid (EDTA) or methanol, which quenches holes in the VB of TiO_2. This H_2 production occurred by the direct electron transfer from the CB of TiO_2 to the active site of hydrogenase at pH range 7–9. This mediator-independent charge transfer is more efficient with *C. pasteurianum* and *D. baculatus* 9974 hydrogenases, and in the presence of methanol as an electron donor with a rate of hydrogen production about 1104 and 1083 μmol H_2 g^{-1} TiO_2 h^{-1}, respectively. Further, the immobilization of rhodium tris- and bis-bipyridyl complexes on the surface of TiO_2 increased the rate of H_2 production in the case of *D. baculatus* 9974 hydrogenase—TiO_2 hybrid to 1422 μmol H_2 g^{-1} TiO_2 h^{-1}. The Rh complex acts as an efficient electron carrier from TiO_2 to the adsorbed hydrogenase enzyme molecules.

For an efficient electron transfer, molecular orientation of the enzymes after immobilizing on the semiconductor should be specifically designed and optimized. A design for electrostatic control of complex formation was achieved by the adsorption of Clostridial [FeFe]-hydrogenase CaI on MPA(3-Mercaptopropionic acid) capped NP-CdTe (Figure 4.11;

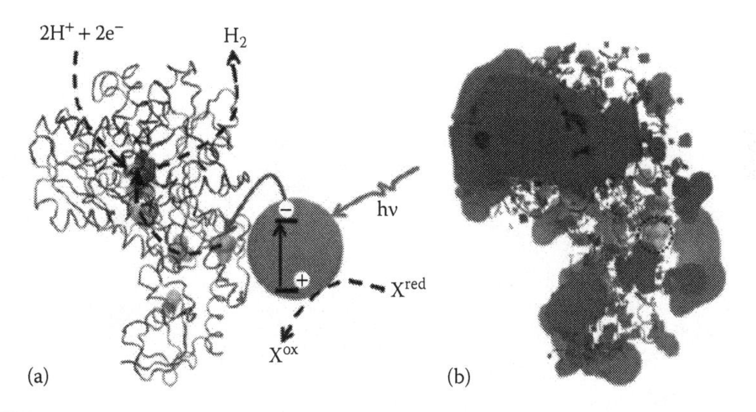

FIGURE 4.11
Functional and structural models for nanocrystal (nc)–CdTe–hydrogenase hybrids. (a) The proposed scheme for light-driven H_2 production by the nc–CdTe:hydrogenase complex. X represents a sacrificial hole scavenger. The hydrogenase structures are homology models of *Clostridium acetobutylicum* [FeFe]-hydrogenase to *C. pasteurianum* [FeFe]-hydrogenase CpI (PDB entry 3C8Y). (b) Electrostatic surface model of the hydrogenase generated using VolMap. Regions of positive and negative charge are shown in light and dark gray color, respectively. (Woolerton, T. W. et al., *Energy and Environmental Science*, 5, 7470–7490, 2012. Reproduced by permission of the Royal Society of Chemistry.)

Brown et al. 2010). The ligand on CdTe binds via thiol group interactions with the CdTe surface, leaving the carboxyl group exposed to solvent. The negative charge developed after the deprotonation of MPA above pH 4.3 leads to the direct binding of hydrogenase in such a way that the protein surface around the distal [4Fe–4S] cluster contains an excess of lysine and arginine residue. While using ascorbic acid as a sacrificial electron donor, the turnover frequency of 25 mol H_2 (mol hydrogenase)$^{-1}$ s^{-1} was obtained but the solar hydrogen production reduced after 5 min of solar irradiation.

For wide band gap semiconductor material, such as TiO_2, a sensitizer complex is required to absorb solar radiation in the visible region. The sensitizer may be a metal-organic framework, for example, ruthenium bipyridyl photosensitizer (Ru II(bpy)$_2$(4,4-(PO_3H_2)$_2$-bpy)]Br_2 ('RuP') (Reisner et al. 2009) or an organic polymeric semiconductor like g-C_3N_4 (Caputo et al. 2015).

Reisner and co-workers immobilized RuP and a *D. baculatum* (Dmb) [NiFeSe]-hydrogenase on P25 Degussa TiO_2 nanoparticles. In this case, TiO_2 acts as an electron transfer mediator. This system has the advantage of avoiding the formation of highly oxidizing holes in the VB of TiO_2 along with extended visible light absorption. The RuP-sensitized system showed a turnover frequency of 50 mol H_2 (mol hydrogenase)$^{-1}$ s^{-1}, under mild conditions (25°C, pH 7.0).

In a similar system, where RuP-TiO_2 was replaced by polyheptazine carbon nitride polymer, the turnover number (TON) of more than 5.8 × 10^5 mol H_2 (mol H_2ase)$^{-1}$ after 72 h using sacrificial electron donor at pH 6 was obtained (Figure 4.12; Caputo et al. 2015). This increased activity is due to the formation

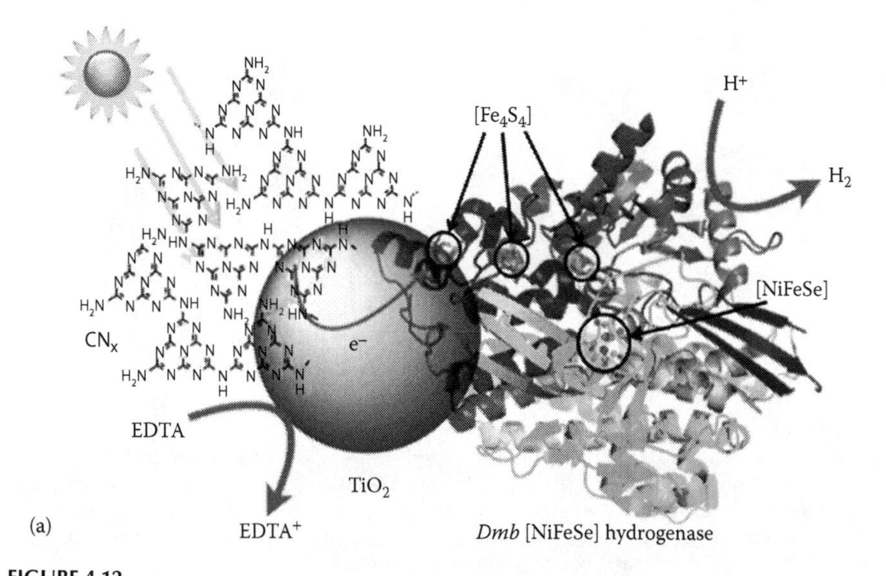

FIGURE 4.12
(a) Schematic representation of photo-H_2 production with Dmb [NiFeSe]-hydrogenase (PDB ID: 1CC1) on CN_x–TiO_2 suspended in water containing EDTA as a hole scavenger. *(Continued)*

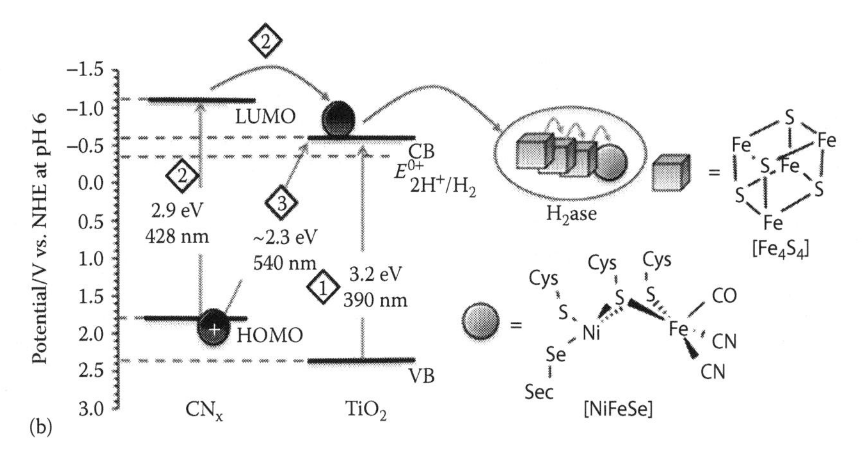

FIGURE 4.12 (Continued)
(b) Irradiation of CN_x–TiO_2 can result in photoinduced electron transfer by three distinct pathways: (1) TiO_2 band gap excitation; (2) excitation of CN_x ($HOMO_{CNx}$–$LUMO_{CNx}$), followed by electron transfer from $LUMO_{CNx}$ into the conduction band of TiO_2 (CB_{TiO2}), (3) charge transfer excitation with direct optical electron transfer from $HOMO_{CNx}$ to CB_{TiO2}. The CB_{TiO2} electrons generated through pathways 1 to 3 are then transferred via the [Fe4S4] clusters to the [NiFeSe] hydrogenase active site. (Caputo, C. A. et al., *Chemical Science*, 6, 5690–5694, 2015. Reproduced by permission of the Royal Society of Chemistry.)

of CN_x–TiO_2 charge transfer complex, facilitating effective electron transfer to hydrogenase and the extended solar radiation absorption.

The above-discussed examples serve an inspiration to design biomimetic photocatalysts. However, the natural enzyme needs to be replaced with their artificial mimics, which can work under normal ambiance conditions and exhibit the catalytic activity as efficient as that of enzymes.

4.7 Conclusion

The strategies and recent progress evolved in designing heterostructured photocatalysts have been discussed. Depending on the direction of the exciton migration from one semiconductor to other, each having different band gap, they are classified into three types. The formation of type II heterojunction has been found to be more promising over the type I heterojunction. The formation of such heterostructured photocatalysts has added advantages, unlike to single band gap photocatalysts, which arise from (1) the synergy of solar radiation harvesting by both the semiconductors (2) in-built electric field at their heterointerface allowing better charge separation of excitons. This phenomenon taking place at the heterointerface lead to overall improved solar H_2 generation efficiency.

Despite the fact that a variety of combinations to design heterostructures have been explored, the design of right heterostructured photocatalysts is under development and yet there are miles to go. The major bottlenecks prevailing in the progress are lower solar-to-hydrogen generation efficiency, poor stability, and the cost of materials being used to design photocatalysts. A careful and exhaustive modeling of the heterointerface, computational studies to understand and predict the band edge offsets, and their optimization are needed. The transient absorption spectroscopy, time-resolved photoluminescence spectroscopy, etc. can be employed extensively to characterize and quantify the separation and transportation of photogenerated charge carriers, which can provide better insight into the intricacies of charge carrier separation at the heterointerface. In order to address the photostability issue, the development of ultra-thin protecting layer that does not compromise the light-harvesting ability requires to be accelerated. Heterostructured photocatalysts show promise to overcome these bottlenecks and could be a reality to generate solar fuels viably in future.

In addition, the design of artificial mimics of enzymes that can work under normal ambiance conditions and exhibit the catalytic activity as efficient as that of enzymes needs to be developed. Such artificial enzyme mimics can be coupled with the semiconductor to form heterostructured photocatalysts.

References

Borgarello, E., J. Kiwi, E. Pelizzatte, M. Visca, and M. Gratzel. 1981. Photochemical cleavage of water by photocatalysis. *Nature* 17, no. 1 (January): 158–160. doi: 10.1038/289158a0.

Brown, K. A., S. Dayal, X. Ai, G. Rumbles, and P. W. King. 2010. Controlled assembly of hydrogenase CdTe nanocrystal hybrids for solar hydrogen production. *Journal of American Chemical Society* 132, no. 28 (June): 9672–9680. doi:10.1021/ja101031r.

Caputo, C. A., L. Wang, R. Beranek, and E. Reisner. 2015. Carbon nitride–TiO_2 hybrid modified with hydrogenase for visible light driven hydrogen production. *Chemical Science* 6, no. 10 (October): 5690–5694. doi:10.1039/C5SC02017D.

Cuendet, P., K. Rao, M. Grätzeland, and D. Hall. 1986. Light induced H_2 evolution in a hydrogenase-TiO_2 particle system by direct electron transfer or via rhodium complexes. *Biochimie* 68, no. 1 (January): 217–221. doi:10.1016/S0300-9084(86)81086-0.

Dinh, T. C., M.-H. Pham, Y. Seo, F. Kleitz, and T.-O. Do. 2014. Design of multicomponent photocatalysts for hydrogen production under visible light using water-soluble titanatenanodiscs. *Nanoscale* 6, no. 9 (January): 4819–4829. doi:10.1039/C3NR06602A.

Fang, J., L. Xu, Z. Zhang, et al. 2013. Au@TiO_2–CdS ternary nanostructures for efficient visible-light-driven hydrogen generation. *ACS Applied Materials & Interfaces* 5, no. 16 (July): 8088–8092. doi:10.1021/am4021654.

Fujishima, A., and K. Honda. 2013. Electrochemical photolysis of water at a semiconductor electrode. *Nature* 238, (July): 37–38. doi:10.1038/238037a0.

Ishida, H., T. Terada, K. Tanaka, and T. Tanaka. 1990. Photochemical carbon dioxide reduction catalyzed by bis(2,2′-bipyridine) dicarbonylruthenium(2+) using triethanolamine and 1-benzyl-1,4-dihydronicotinamide as an electron donor. *Inorganic Chemistry* 29, no. 5 (March): 905–911. doi:10.1021/ic00330a004.

Kuang, Y., Q. Jia, H. Nishiyama et al. 2016. A front-illuminated nanostructured transparent BiVO4 photoanode for >2% efficient water splitting. *Advance Energy Materials* 6, no. 2 (January): 1501645. doi:10.1002/aenm.201501645.

Lakadamyali, F., A. Renal, M. Kato, J. R. Durrant, and E. Reisner. 2012. Electron transfer in dye-sensitised semiconductors modified with molecular Cobalt Catalysts: photoreduction of aqueous protons. *Chemistry* 18, no. 48 (November): 15464–15475. doi:10.1002/chem.201202149.

Li, J., and N. Wu. 2015. Semiconductor-based photocatalysts and photoelectrochemical cells for solar fuel generation: A review. *Catalysis Science & Technology* 5, no. 3 (January): 1360–1384. doi:10.1039/C4CY00974F.

Marschall, R. 2014. Semiconductor composites: strategies for enhancing charge carrier separation to improve photocatalytic activity. *Advanced Functional Materials* 24, no. 17 (May): 2421–2440. doi:10.1002/adfm.201470108.

May M. M., H. Lewerenz, D. Lackner, F. Dimroth and T. Hannappel. 2015. Efficient direct solar-to-hydrogen conversion by in situ interface transformation of a tandem structure. *Nature Communications* 6, (September): 1–7. doi:10.1038/ncomms9286.

Nanda, K. K., S. Swain, B. Satpati, L. Besra, B. Mishra, and Y. S. Chaudhary. 2015. Enhanced photocatalytic activity and charge Carrier dynamics of hetero-Structured organic–inorganic nano-photocatalysts. *ACS Applied Materials & Interfaces* 7, no. 15 (March): 7970–7978. doi:10.1021/acsami.5b00022.

Nanda, K. K., S. Swain, B. Satpati, L. Besra, and Y. S. Chaudhary. 2014. Facile synthesis and the photo-catalytic behavior of core–shell nanorods. *RSC Advances* 4, no. 21 (January): 10928–10934. doi:10.1039/c3ra47024e.

Ng, Y. H., A. Iwase, A. Kudo, and R. Amal. 2010. Reducing graphene oxide on a visible-light $BiVO_4$ photocatalyst for an enhanced photoelectrochemical water splitting. *Journal of Physical Chemistry Letters* 1, no. 17 (August): 2607–2612. doi:10.1021/jz100978u.

Reisner, E., D. J. Powell, C. Cavazza, J. C. Fontecilla-Camps, and F. A. Armstrong. 2009. Visible light-driven H_2 production by hydrogenases attached to dye-sensitized TiO_2 nanoparticles. *Journal of the American Chemical Society* 131, no. 51 (November): 18457–18466. doi:10.1021/ja907923r.

Reisner, E., J. C. Fontecilla-Campsb, and F. A. Armstrong. 2009. Catalytic electrochemistry of a [NiFeSe]-hydrogenase on TiO_2 and demonstration of its suitability for visible-light driven H_2 production. *Chemical Communications* 2009, no. 5 (February): 550–552. doi:10.1039/B817371K.

Robel, I., M. Kuno, and P. V. Kamat. 2007. Size-dependent electron injection from excited CdSe quantum dots into TiO_2 nanoparticles. *Journal of American Chemical Society* 129, no. 14 (March): 4136–4137. doi:10.1021/ja070099a.

Sasaki, Y., H. Nemoto, K. Saito, and A. Kudo. 2009. Solar water splitting using powdered photocatalysts driven by Z-schematic inter-particle electron transfer without an electron mediator. *Journal of Physical Chemistry C* 113, no. 40 (September): 17536–17542. doi:10.1021/jp907128k.

Sayama, K., K. Mukasa, R. Abe, Y. Abe, and H. Arakawa. 2002. A new photocatalytic water splitting system under visible light irradiation mimicking a Z-scheme mechanism in photosynthesis. *Journal of Photochemistry and Photobiology A: Chemistry* 148, no. 1–3 (May): 71–77. doi:10.1016/S1010-6030(02)00070-9.

Song, X. W., H. M. Wen, C. B. Ma, H. H. Cui, H. Chen, and C. N. Chen. 2014. Efficient photocatalytic hydrogen evolution with end-group-functionalized cobaloxime catalysts in combination with graphite-like C3N4. *RSC Advances* 4, no. 36 (April): 18853–18861. doi:10.1039/c4ra01413h.

Svetlitchnyi, V., C. Peschel, G. Acker, and O. Meyer. 2001. Two membrane-associated NiFeS-carbon monoxide dehydrogenases from the anaerobic carbon-monoxide-utilizing Eubacterium Carboxydothermushydrogenoformans. *Journal of Bacteriology* 183, no. 17 (September): 5134–5144. doi:10.1128/JB.183.17.5134-5144.2001.

Vignais, P. M., and B. Billoud. 2007. Occurrence, classification, and biological function of hydrogenases: an overview. *Chemical Review* 107, no. 10 (October): 4206–4272. doi:10.1021/cr050196r.

Wang, X., G. Liu, L. Wang, Z.-G. Chen, G. Q. (Max) Lu, and H.-M. Cheng. 2012. ZnO–CdS@Cdheterostructure for effective photocatalytic hydrogen generation. *Advanced Energy Materials* 2, no. 1 (January): 42–46. doi:10.1002/aenm.201100528.

Wang, X., G. Liu, Z. G. Chen, et al. 2009. Enhanced photocatalytic hydrogen evolution by prolonging the lifetime of carriers in ZnO/CdS heterostructures. *Chemical Communications*, no. 23 (May): 3452–3454. doi:10.1039/B904668B.

Woolerton, T. W., S. Sheard, Y. S. Chaudhary, and F. A. Armstrong. 2012. Enzymes and bio-inspired electrocatalysts in solar fuel devices. *Energy and Environmental Science* 5, no. 16 (June): 7470–7490. doi:10.1039/C2EE21471G.

Xie, M., X. Fu, L. Jing, P. Luan, Y. Feng, and H. Fu. 2014. Long-lived, visible-light-excited charge carriers of $TiO_2/BiVO_4$ nanocomposites and their unexpected photoactivity for water splitting. *Advanced Energy Materials* 4, no. 5 (April): 1300995–1301001. doi:10.1002/aenm.201300995.

Xinga, Z., X. Zonga, Y. Zhua, Z. G. Chenb, Y. Baia, and L. Wanga. 2015. A nanohybrid of CdTe@CdS nanocrystals and titaniananosheets with p-n nanojunctions for improved visible light-driven H_2 production. *Catalysis Today* 264: 229–235. doi:10.1016/j.cattod.2015.08.007.

Xu, M., T. Ye, F. Dai, et al. 2015. Rationally designed n–n heterojunction with highly efficient solar hydrogen evolution. *ChemSusChem* 8, no. 7 (April): 1218–1225. doi:10.1002/cssc.201403334.

Yu, J. G., S. H. Wang, J. X. Low, and W. Xiao. 2013a. Enhanced photocatalytic performance of direct Z-scheme g-C3N4–TiO2 photocatalysts for the decomposition of formaldehyde in air. *Physical Chemistry Chemical Physics* 15, no. 39 (August): 16883–16890. doi:10.1039/C3CP53131G.

Yu, W., D. Noureldine, T. Isimjan, et al. 2015. Nano-design of quantum dot-based photocatalysts for hydrogen generation using advanced surface molecular chemistry. *Physical Chemistry Chemical Physics* 17, no. 2 (November): 1001–1009. doi:10.1039/C4CP04365K.

Yu, Z. B., Y. P. Xie, G. Liu, G. Q. Lu, X. L. Ma, and H. M. Cheng. 2013b. Self-assembled CdS/Au/ZnO heterostructure induced by surface polar charges for efficient photocatalytic hydrogen evolution. *Journal of Materials Chemistry A* 1, no. 8 (January): 2773–2776. doi:10.1039/C3TA01476B.

Yuan, Y., L. W. Ruan, J. Barber, S. Chye, J. Loo, and C. Xuec. 2014. Hetero-nanostructured suspended photocatalysts for solar-to-fuel conversion. *Energy & Environmental Science* 7, no. 12 (October): 3934–3951. doi:10.1039/C4EE02914C.

Zhang, J., M. Zhang, R. Q. Sun, and X. Wang. 2012. A facile band alignment of polymeric carbon nitride semiconductors to construct isotype heterojunctions. *Angewandte Chemie International Edition* 51, no. 40 (October): 10145–10149. doi:10.1002/anie.201205333.

Zhang, X., B. Peng, S. Zhang, and T. P. Robust. 2015. Robust wide visible-light-responsive photoactivity for H_2 production over a polymer/polymer heterojunction photocatalyst: the significance of sacrificial reagent. *ACS Sustainable Chemistry & Engineering* 3, no. 7 (June): 1501–1509. doi:10.1021/acssuschemeng.5b00211.

Zhang, X., Z. Meng, and D. Rao, et al. 2016. Efficient band structure tuning, charge separation, and visible-light response in ZrS_2-based van der Waals hetero-structures. *Energy & Environmental Science* 9, no. 3 (January): 841–849. doi:10.1039/C5EE03490F.

Zhou, P., J. Yu, and M. Jaroniec. 2014. All-solid-state Z-scheme photocatalytic systems. *Advanced Materials* 26, no. 29 (August): 4920–4935. doi:10.1002/adma.201400288.

5

Thermochemical Hydrogen Generation

Randhir Singh and Debasis Saran

CONTENTS

5.1 Introduction

Hydrogen does not occur in the elemental form in nature. Currently about 95% of hydrogen is produced from hydrocarbons—the most economic route to make hydrogen. The remaining supply of hydrogen (~5%) is met by electrolysis of water—the most abundant source of hydrogen. In view of the accelerated depletion of the fossil fuels and the CO_2-induced global warming, the humanity has to find a new and renewable source of hydrogen. A thermochemical route of hydrogen (H_2) production employs heat or a combination of heat and O_2/water to extract hydrogen bound to a fuel or water.

A cyclic thermochemical route of hydrogen (H_2) production requires as net input to the process only heat and water or H_2S (Kappauf and Fletcher 1989). A "hybrid" thermochemical cycle of H_2 production uses, in addition, electricity. The overall reaction of the process, hybrid or not, is to extract the hydrogen atoms from the water molecule in a separate stream than the oxygen. Figure 5.1 gives an overview of the various thermochemical processes, cyclic or noncyclic, commercial, or under research, to produce the H_2.

The easiest way of producing H_2 might appear to thermally decompose the water into its constituents, that is, into hydrogen and oxygen, and to separate this hot gas mixture before cooling it down to avoid the recombination. Another strategy to separate the H_2 and O_2 gas mixture is to first quench it from the decomposition temperature and then separate the mixture using Pd membrane. However, the water decomposition reaction

$$H_2O(g) = H_2(g) + \frac{1}{2}O_2(g) \quad (5.1)$$

is not spontaneous and has a positive free energy change ($\Delta G^0_{298.15}$) of 228.58 kJ/mol at 298.15 K and 1 atm pressure. The variation of free energy of formation of water (g) with temperature is shown in Figure 5.2 (Chase 1998). To make the water decomposition reaction spontaneous ($G^0_T = 0$), the temperature required would be over 4325 K! Although reaction (5.1) can be driven to appreciable degree by using nonstandard conditions such as a reduced pressure or an inert atmosphere at lower temperature of about 2500 K, the temperature is still impracticably high (Baykara 2004; Nakamura 1977).

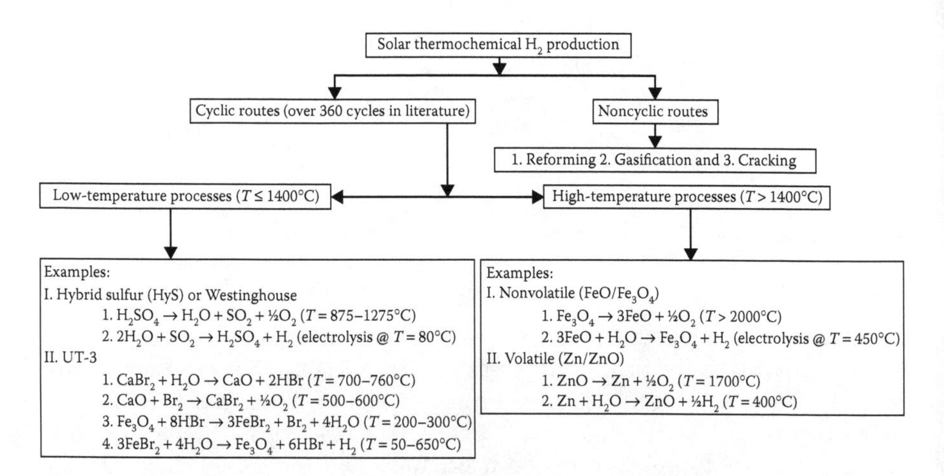

FIGURE 5.1
An overview of the various thermochemical processes, cyclic or noncyclic, commercial, or under research, to produce H_2.

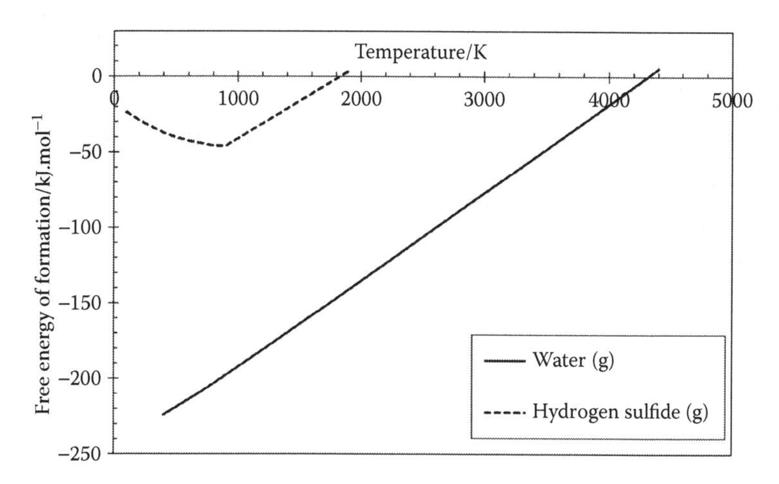

FIGURE 5.2
Variation of free energy of formation with temperature. (Data from Chase, M. W., *Journal of Physical and Chemical Reference Data Monograph No. 9*, American Institute of Physics.)

To overcome the barrier of high temperature in the one-step water thermolysis, a multistep process can theoretically be envisaged wherein the process steps are carried out at different temperatures, each at much lower than 2000 K, so that the sum total of all process steps is reaction (5.1). The multistep thermochemical cycles in one form of their manifestation employ a reversible redox mediator that in the first step gets oxidized from the oxygen in water producing H_2 (hydrolysis reaction) at a temperature T_1 and in the subsequent step, say at temperature T_2 ($>T_1$), the oxidized form of the mediator is thermally decomposed back to its active state, which is ready to undergo the H_2-producing hydrolysis reaction of step 1. The two of the reactions are repeated cyclically, as shown in Figure 5.3, for an iron-based redox mediator. Thus, a multistep thermochemical process needs, in addition to heat and water, a redox mediator as a material input that cycles during the process between different oxidation states but is not consumed.

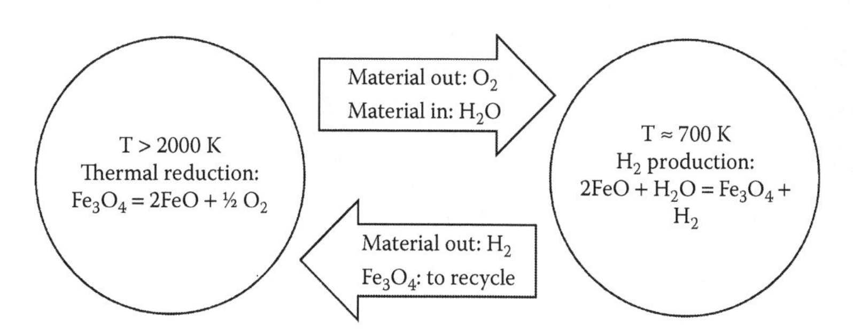

FIGURE 5.3
A schematic of a two-step thermochemical process to produce hydrogen.

Some of the thermochemical routes to hydrogen production are noncyclic in nature and require a fossil-based or renewably made methane (CH_4) from biomass as a consumable material input and solar heat as an energy input. Examples of such processes are steam methane reforming, coal reforming, biomass gasification, biofuel such as bioethanol reforming, and aqueous phase reforming of bioderived carbohydrates (Funk 2001). These noncyclic thermochemical pathways employ carbonaceous material to produce hydrogen. The carbon in these compounds gets converted into CO_2. Obviously, in the absence of any carbon capture and storage (CCS), these processes will emit large amounts of CO_2. For example, 1 kg of H_2 production from natural gas (NG) and coal generates 7.3 and 29 kg of CO_2, respectively (Kothari et al. 2008). Notwithstanding the CO_2 emission, H_2 generation from reforming of the fossil-based fuels is going to linger for at least the short term, in the run up to 2020 (Pregger et al. 2009). Nonthermochemical routes to H_2 generation from water are water electrolysis, photoconversions (photoelectrochemical, photocatalytic, and photobiological), and fermentation. The projected timeline for the various commercial-scale H_2 production pathways in United States is shown in Table 5.1 (Anon 2015). The same trend is expected to be followed throughout the developed world receiving useable amount of solar insolation. Developing countries are expected to follow the same trend with a time lag of 5–10 years.

5.1.1 Hydrogen: Availability, Imperative, and Applications

Hydrogen is the fifth most abundant element in terms of atom fraction in the upper continental Earth's crust after O, Si, Al, and Na (Haxel and Hedrick 2002). It is so reactive as to occur in nature only in the form of a compound with elements such as oxygen and carbon. According to International Energy Outlook 2013 (2013), the world energy consumption will grow by 56% by 2040 from its 2010 level, with about 80% of the demand met by fossil fuels (International Energy Outlook 2013). The depletion times for oil, coal, and gas have been estimated as approximately 35, 107, and 37 years, respectively, in 2009 (Shafiee and Topal 2009). Add to

TABLE 5.1

Projected Timeline for the Various Commercial Scale H_2 Production Pathways in USA

Timeline	Technologies
Near term (current industrial methods)	Natural gas reforming, Biomass gasification
Midterm (biomass pathways)	Biomass gasification, coal gasification with carbon capture and storage (CCS), Electrolysis (wind)
Long term (Solar Pathways)	High-temperature electrolysis, solar thermochemical hydrogen (STCH), photoelectrochemical (PEC), photobiological, electrolysis (solar)

Source: Anon, http://energy.gov/eere/fuelcells/hydrogen-production-pathways, 2015.

this the deteriorating urban air pollution level due to the usage of fossil fuels and a threat to global warming accompanied by a climate change, the development and use of alternative, clean energy sources cannot be overemphasized. When it comes to a realistic clean and efficient energy vector, nothing can possibly beat hydrogen (Scott 2008). Unlike fossil fuels, burning hydrogen, or its usage in a fuel cell to generate electricity, merely produces water. Hydrogen is projected to be the most important energy vector of the future.

Currently, hydrogen finds many industrial usages as shown in Figure 5.4 (Abbasi and Abbasi 2011; Ramachandran and Menon 1998). The largest consumer of H_2 is the ammonia plant. It is expected that the transport sector is going to emerge as one of the largest consumers of H_2 in future. The H_2 demand for the transportation needs in the United States alone is projected to be about 150 million tons per year (Dresselhaus et al. 2003; Turner 2004).

5.1.2 Hydrogen Economy: Historical Perspective

Post–World War II, there was a short-lived optimism in the United States lasting until the 1973 oil embargo that the electricity from nuclear power plants would be so cheap as to obviate the need of metering (Fletcher 2001). The U.S. Army and others carried out a project in the early 1960s, Energy Depot, with an objective to produce a fuel from air, water, and portable nuclear heat; the upshots of the project were not satisfactory though (Funk 2001). The imposition of the 1973 oil embargo triggered a flurry of activities to find an alternative to fossil fuels, especially the generation of hydrogen

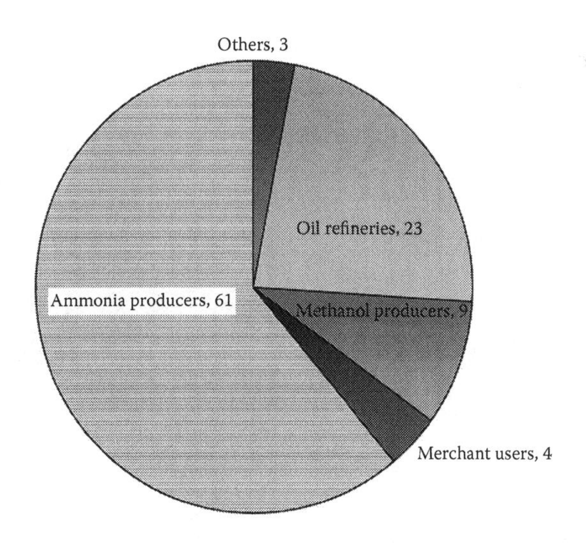

FIGURE 5.4
The industrial usage of H_2 in various sectors. (After Abbasi, T., and S. A. Abbasi, *Renewable and Sustainable Energy Reviews*, 15(6), 3034–40, 2011.)

from nuclear heat and water. The intensive research effort lasted until about 1985 and stayed tepid over the ensuing period of 1985–2000. However, the international consensus toward an emerging hydrogen economy—where the role of hydrocarbon fuel in current economy is envisioned to be replaced by the hydrogen sans greenhouse gas emissions—has been growing consistently since the oil crisis. Many nations and international groups such as the United States, Japan, European Union, India, among others have charted out elaborate plans to make a smooth transition toward the hydrogen economy. In fact, the global research output on "thermochemical hydrogen production" has been gaining momentum since the turn of the twenty-first century, as is clear from Figure 5.5. It is to be noted, however, that the vision for the hydrogen economy goes much further back. In 1874, H_2 was termed "the coal of the future" by Jules Verne. Likewise, H_2 was proposed as the transportation fuel as early as the 1930s (Turner 2004).

5.1.3 Thermodynamics of Thermochemical H_2 Production

Understanding of thermodynamics principles plays an important role for efficient operation of a thermochemical cycle. The maximum chemical free energy stored in fuel generated thermochemically is limited by the maximum efficiency of a heat engine, that is, the Carnot limitation (Fletcher and Moen 1977). A Carnot heat engine is basically a cyclic device that takes heat energy from high-temperature source, does some useful work, and rejects

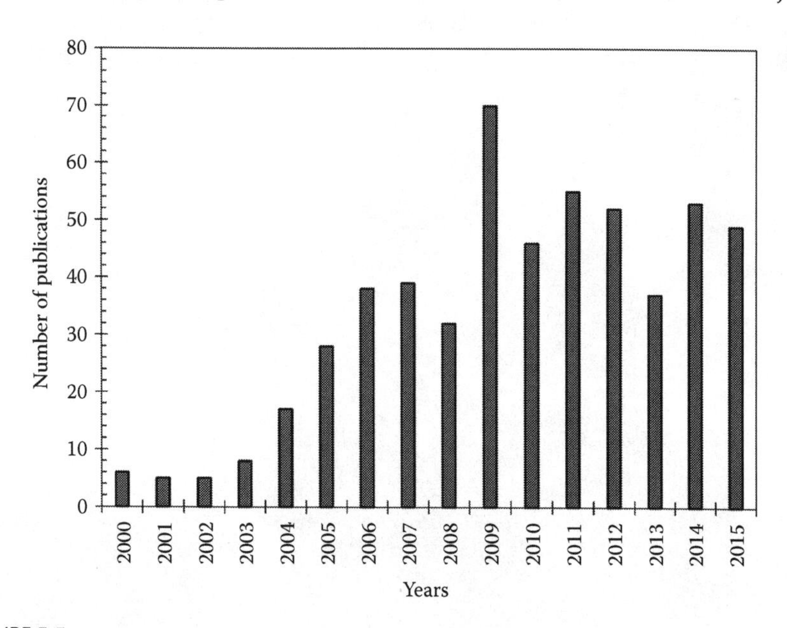

FIGURE 5.5
The global research output on "thermochemical hydrogen production" since 2001 measured by the number of publications in the Scopus.

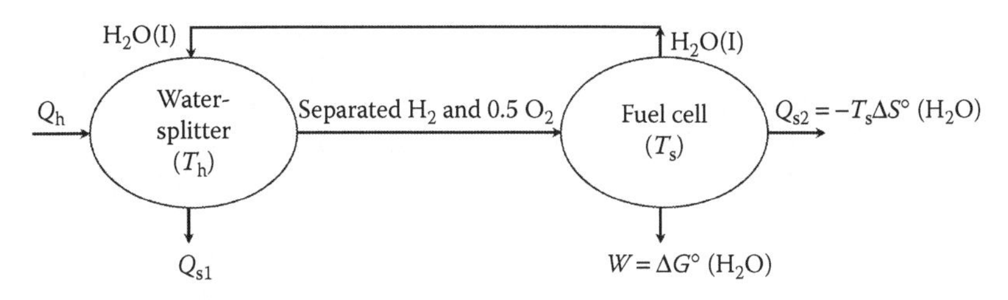

FIGURE 5.6
Idealization of a thermochemical cycle to produce hydrogen and a fuel cell as a reversible Carnot heat engine. (After Fletcher, E. A., *Journal of Solar Energy Engineering*, 123(2), 63–74, 2001.)

rest of it to the surroundings at low temperature. A similar analogy can be drawn from a solar thermochemical water splitting cycle by visualizing a fuel cell along with it (Fletcher 2001).

Consider Figure 5.6 for an idealization of a cyclic water splitting reactor. Water at ambient conditions is fed into a thermochemical water splitter, which exchanges heat with two reservoirs at temperatures of T_h and T_S ($T_h > T_S$), respectively. The splitter produces H_2 and O_2 in separate streams from the net material input of water and a net thermal input of ($Q_h - Q_S$) at a temperature of T_h. Thus, the water splitter effects the chemical reaction:

$$H_2O\ (l, 1\ \text{bar}, 298.15\ \text{K}) = H_2\ (g, 1\ \text{bar}, 298.15\ \text{K}) + 0.5\ O_2\ (g, 1\ \text{bar}, 298.15\ \text{K}) \tag{5.2}$$

Reaction (5.2) is endothermic, having a standard enthalpy change of 285.15 kJ/mol at 298.15 K. This reaction could proceed via a multistep process, each carried out at a different temperature; the temperature T_h refers to the maximum temperature across all process steps. At 298.15 K, we must provide the useful energy ($\Delta G°$) in addition to a minimum amount of heat equal to $T\Delta S°$, where $\Delta G°$ and $\Delta S°$ are 237.14 kJ/mol and 0.16 kJ/mol K, respectively, at 298.15 K. However, supplying 237.14 kJ/mol of electrical energy ($\Delta G°$) for decomposing just 1 mol of water is not an energy-efficient choice, as we need around 700 kJ from our prime thermal energy input, which is 2.5 times the $\Delta H°$ at 298.15 K. The pure heat and electrical energy requirement for reaction (5.2) will change as a function of the process temperature; however, at T_h = 298.15 K, there is no way but to expend 237.14 kJ of electrical energy, which is not the most viable economic option for generation of H_2 gas, except in cases where cheap renewable electricity is available in plenty. Back conversion of H_2 and O_2 into H_2O in a device such as a hypothetical reversible fuel cell would produce the same amount of useful work and heat. It is interesting to see what happens when the water splitter is operated at a temperature $T_h >> T_S$. In particular, what is the theoretical maximum useful work that can be performed by the separate H_2 and O_2 outlet streams from the splitter? The answer to this question is the work from a reversible Carnot engine operating between T_h and T_S. If the useful work in the derived fuel were any greater than this limit, the second law of thermodynamics would be in violation. This is a very useful conclusion applicable to all thermofuel production processes.

It is the concentrated solar energy that is deemed to be the heat source at the high temperature T_h. The maximum source temperature attainable by using a solar energy concentrating device on Earth is about 5800 K, equivalent to Sun's surface temperature. The sink temperature can be taken as Earth's surrounding temperature of 298.15 K. In accordance with Carnot's principle, a heat engine operating at these source and sink temperatures will convert about [1 − (298.15/5800)] × 100 = 94.86% of solar energy into useful work. At a source temperature lower than 5800 K, the reversible Carnot efficiency—the maximum attainable efficiency in a reactor—drops monotonically to 85% and 69.8% at temperatures of 2000 and 1000 K, respectively, as shown in Figure 5.7.

Now, we are in a position to analyze the system in a more general way. The products that are coming out of the splitter, that is, H_2 and O_2 are fed into the fuel cell, which, in turn, produces liquid water and does electrical work of $\Delta G_{298.15K}$ of 237.14 kJ for each mol of water split. The water produced by the fuel cell is fed back to the splitter to complete the cycle. If the water splitter

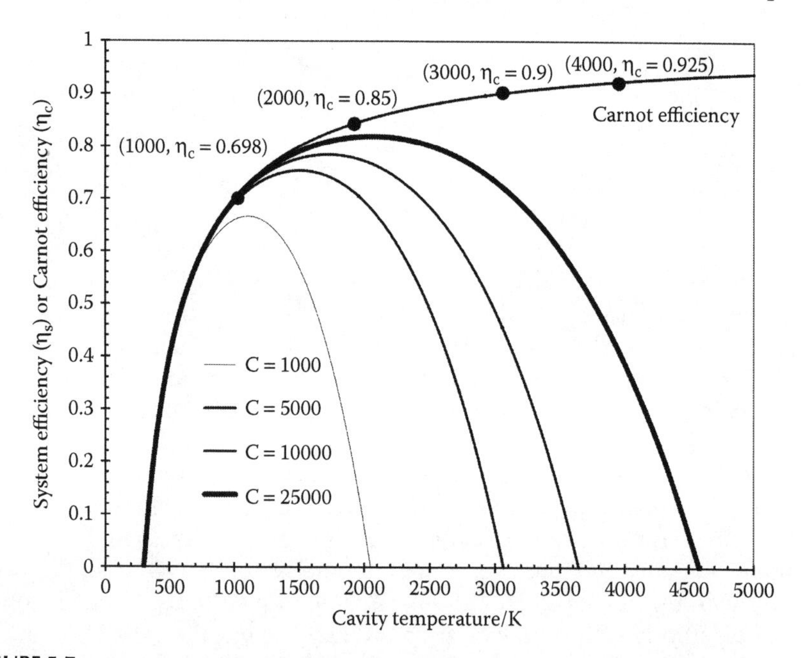

FIGURE 5.7
Variation of the ideal system efficiency with cavity temperature at various solar concentrations. (After Fletcher, E. A., *Journal of Solar Energy Engineering*, 123(2), 63–74, 2001.)

and fuel cell were the parts of a single enclosure, then this enclosure would act as a heat engine whose maximum efficiency is limited by the Carnot efficiency.

A thermochemical reaction can be carried out using solar energy, which, at first, is concentrated to obtain a desired temperature. The concentrated heat and the reactants meet in a *reactor* or in a *receiver.* A schematic of a thermochemical reactor operating on solar energy is shown in Figure 5.8. A solar *collector* captures incident Sun radiation and then focuses it to a solar receiver/reactor, which is placed at the focal plane of the collector. Three main solar collector design configurations used nowadays are trough, tower, and dish systems. These configurations use different methods to focus Sun's rays onto a receiver placed at the focal plane. In a trough system, linear, parabolic mirrors focus solar insolation onto a tube-like receiver placed along the focal line. A tower system uses a field of tracked parabolic mirrors, which focus the solar radiation onto a solar receiver mounted on a nearby tower. Finally, in a dish system, the solar radiation is focused on a solar receiver at its focus using a paraboloidal mirror.

Let us now define some terminologies in solar reactor engineering. The *absorption* efficiency of a cavity receiver, η_A, is defined as the fraction of the energy incident on the solar concentrator that is absorbed after accounting for the losses through inefficient optics and emission via cavity aperture. It is given by

$$\eta_A = \frac{IA\eta_0\alpha - a\varepsilon\sigma T_h^4}{IA} \tag{5.3}$$

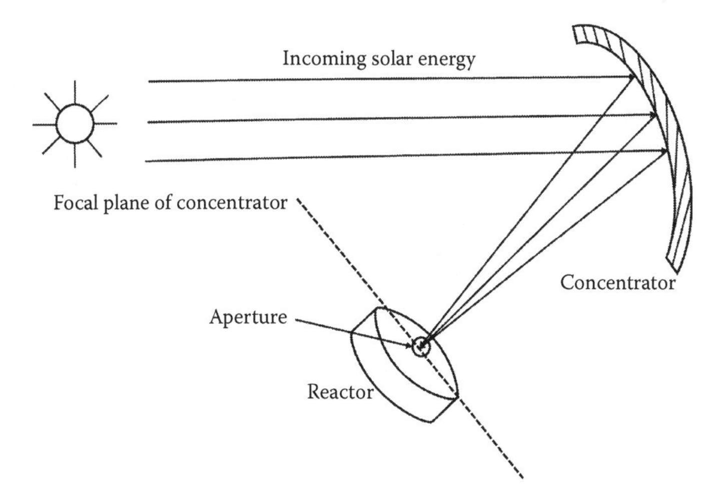

FIGURE 5.8
A schematic of a thermochemical reactor operating on solar energy.

where:

I is the intensity of the incident sunlight
A is the area of the collector normal to the direction of the sunlight
a is the reradiating area, that is, the area of the aperture of the cavity
η_0 is the efficiency that accounts for energy losses due to imperfections of the optics
α is the absorptance
ε is the emittance of the receiver
σ is the Stefan Boltzmann constant
T_h is the temperature in the cavity

With perfect optics and absorptance and emittance of 1, Equation (5.3) reduces to

$$\eta_A = \frac{(IC)-(\sigma T_h^4)}{IC} \tag{5.4}$$

where:

C is the solar concentration ratio defined as Q_{solar}/IA, where Q_{solar} is the solar power intercepted by the concentrator

Higher concentration ratio signifies high concentrating capability of collection systems to collect sunlight. The solar flux concentration ratio typically obtained is at the level of 100, 1000, and 10,000 suns (1 Sun = 1 kW/m^2) for trough, tower, and dish systems, respectively.

The variation of the absorption efficiency of a black body cavity at various solar concentration ratio is plotted as a function of its temperature in Figure 5.9. We can easily conclude that as we increase the concentration ratio, the maximum temperature in the cavity increases accordingly. The maximum temperature at a fixed C is attained when η_A in Equation (4) goes to zero. For example, for a concentration ratio of 1000, the absorption efficiency is 0.9433 at a cavity temperature of 1000 K, and we can achieve a maximum temperature of about 2050 K.

Efficiency of any system is defined as the ratio of useful work output to input energy. The endothermic reaction (5.2) when carried out at T_h requires no or little energy. However, reaction (5.2) in reverse at T_S when carried out in a fuel cell-like device produces useful work equal to the $\Delta G°$ of water formation at T_S. Now, we are in a position to define system efficiency, η_S—the ratio of useful work produced by reaction (5.2) in reverse to the solar energy intercepted by the concentrator. It is also the product of the absorption efficiency, η_A and the Carnot efficiency (η_C) given by

$$\eta_S = \frac{(IA-\sigma T_h^4)}{IA} \times \frac{T_h - T_S}{T_h} \tag{5.5}$$

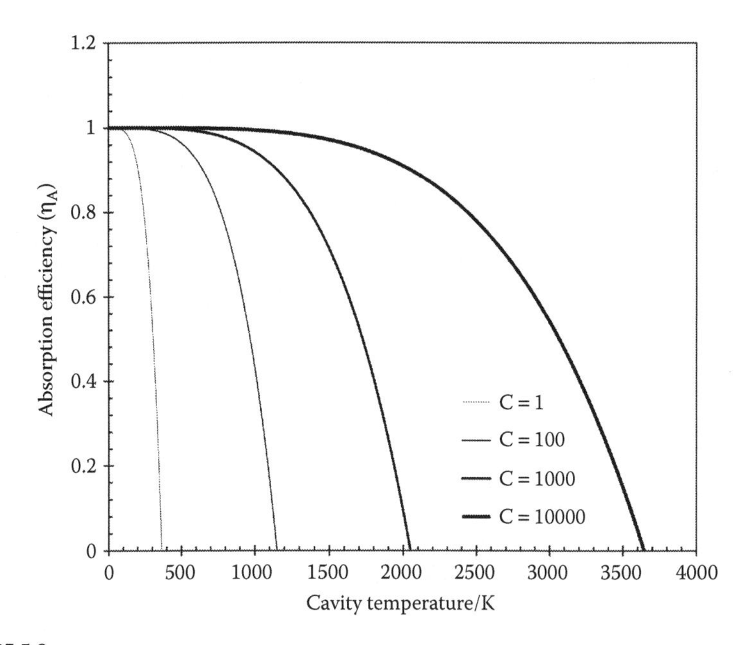

FIGURE 5.9
Variation of the energy absorption efficiency of a black body cavity with its temperature at various solar concentration ratios. (After Fletcher, E. A., *Journal of Solar Energy Engineering*, 123(2), 63–74, 2001.)

where each term has the same meaning as discussed above and T_S is the sink temperature and the system optics is assumed to be perfect.

Figure 5.7 shows the variation of the system efficiency with cavity temperature when the ambient temperature is 300 K. Carnot efficiency and absorption efficiency, η_A, both are contrasting in nature. Carnot efficiency (η_C) increases monotonically, whereas absorption efficiency decreases with T_h. In Figure 5.7, we can observe that system efficiency, η_S, which is the product of above two efficiencies, follows the pattern of both. When we start increasing the cavity temperature, system efficiency first rises to attain a maximum value, then decreases as cavity temperature increases, and finally settles to zero at the *stagnant* temperature. It is imperative that we must carry out the reaction at optimum temperature, that is, the temperature where η_S is maximum without compromising operating conditions and durability of materials. From Figures 5.9 and 5.7, it is clear that with an increase in concentration ratio, both absorption efficiency and maximum system efficiency increase. So, it is essential that we focus on increasing the quality of optical devices to fully utilize Sun's temperature of 5800 K for the maximum gain in η_S.

5.1.4 Scope

Solar thermochemistry has grown enormously over the past few decades. Within this, the thermochemical hydrogen production assumes a prominent field of active research these days. This chapter introduces the different routes to thermochemical hydrogen production from renewable (water and biomass) as well as nonrenewable sources (fossil-based fuel, H_2S). The authors have attempted to incorporate most recent developments in this area. Emphasis has also been placed on simple and cogent description of various aspects of the thermochemical hydrogen production. In the following section, select thermochemical approaches with a promise of high energy efficiency have been classified, described, and compared. Then, in Section 5.3, the state of the art in *cyclic* thermochemical hydrogen production processes is presented with regard to material and reactor technologies. Finally, a summary and an outlook are put in Section 5.4.

5.2 Thermochemical Approaches

There are primarily three fundamental ways to thermochemical hydrogen production (Steinfeld 2005): (1) solar thermolysis of water (*one-step* process); (2) thermochemical cycles to decompose water in a *multistep* process, each step being usually carried out at a different temperature; and (3) decomposition of fossil fuels, biomass, or biomass-derived fuels such as methane. Various technologies, each relying on a heavily endothermic step of decomposing the hydrogen-containing molecule/s, are summarized in Table 5.2. The process heat requirement can either be met by nuclear or by burning a portion of the precursor fuel for process

TABLE 5.2

Technologies for Thermochemical Production of H_2 or H_2-Containing Fuel. A Separation Step is Needed to Extract H_2 from the H_2-Containing Fuels, which Are Obtained from Reforming, Gasification, and Cracking of a Carbonaceous Fuel

Sl. No.	Thermochemical Processes	Process Reactions
1.	Thermolysis	1. $H_2O = H_2 + 0.5\ O_2$ 2. $H_2S = H_2 + S$
2.	Thermochemical cycles	Over 360 types of cycles based on the cycling material (Heske et al. 2011)! For an iron-based cycle, the reactions are: 1. $Fe_3O_4 \rightarrow 3FeO + ½O_2$ ($T > 2000°C$) 2. $3FeO + H_2O \rightarrow Fe_3O_4 + H_2$ (electrolysis @$T = 450°C$)
3.	High-temperature electrolysis	1. $H_2O + 2e- \rightarrow H_2 + O_2-$ (cathode) 2. $O_2- \rightarrow ½\ O_2 + 2e-$ (anode) Overall: $H_2O \rightarrow H_2 + ½\ O_2$
4.	Reforming	$C_xH_yO_z + (x-z)H_2O \rightarrow \{y/2 + (x-z)\}.H_2 + xCO$
5.	Gasification	$C_xH_yO_z + (x-z)H_2O \rightarrow \{y/2 + (x-z)\}.H_2 + xCO$
6.	Cracking	$C_xH_y \rightarrow xC\ (gr) + y/2\ H_2$

temperature less than 1223 K (Roeb et al. 2012) or concentrated solar heat for temperatures up to or under ~2223 K, the maximum temperature being limited by the material durability rather than the attainable temperature in a receiver reactor.

Except for the high-temperature electrolysis (HTE), the rest of the processes to thermochemically produce hydrogen may result in a mixture of gas containing H_2, necessitating a *separation* stage. The separation step can be a very complicated process especially at high temperature such as the thermolysis of water, the Zn–ZnO-based thermochemical cycles. The derivation of hydrogen from a lean fuel such as biomass or a polluting fuel such as coal via *reforming*, *gasification*, and *cracking* results in an H_2-containg mixture with gases CO, CH_4, CO_2, and so on, which can be separated by a variety of well-established commercial techniques at relatively low temperature. In the following subsections, individual processes are explained in detail.

5.2.1 Thermolysis (Solar)

The naturally occurring source materials for thermolysis are water and H_2S, requiring a reversible decomposition temperature of, respectively, 4300 and 1800 K, as is clear from Figure 5.2.

Apparently, only water has been tried in experimental solar reactors at Montreal, Canada, at temperatures of 2000 and 2500 K (Baykara 2004). As is obvious from the direct thermolysis reaction (5.1), the product of the reaction is a mixture of H_2 and O_2 or a very active mix of radicals involving atomic hydrogen or oxygen, which becomes an explosive mixture at lower temperatures. Thus, the product gases need to be either separated at the decomposition temperature or be quenched so fast that the recombination reaction [i.e., the reverse of reaction (5.1)] does not occur appreciably. Typical cooling rate required to suppress the back reaction is over 10^5 K/s, achievable with some cold diluting gas that makes the process of cold separation of the mixture more difficult. Moreover, the thermodynamic limit of conversion of water at 2500 K is merely 4% and the thermal efficiency hardly exceeded 1% in experimental solar reactor. Solar water thermolysis, although actively investigated today, is marred by the requirement of higher and higher temperature to achieve an appreciable degree of decomposition and by the need for a high-temperature separation to increase the process efficiency. Naturally, the lack of a durable reactor design and a suitable construction material are major stumbling blocks in making this process viable.

An attractive alternative to high-temperature thermolysis is comparatively low-temperature (1600 K) *catalytic* decomposition of water on Pt (Jellinek and Kachi 1984; Tsai et al. 2016). The extent of decomposition is subject to the same thermodynamic limit applicable to high-temperature thermolysis. However, the higher conversion rate and simplified gas separation protocol without a quench make the process promising.

5.2.2 Thermochemical Cycles

A large-scale project on "Solar Hydrogen Generation Research" from the Department of Energy, USA, was carried out by a consortium of US universities and national laboratories. The project concluded in June 2011 and the final report was made available publically (Heske et al. 2011), with a *go/no-go* decision to be made based on economic potential as to whether or not to proceed to pilot plant design. Of the two tasks under the project, one was dedicated on Solar Thermochemical Hydrogen Production (STCH). Under the STCH task, a total of 360 thermochemical cycles to produce hydrogen were ranked based on a number of well-defined quantifiable objective criterion, each contributing to a maximum (total) score of 100. The top 67 cycles were moved to next phase of selection. Finally, the set of thermochemical cycles that appeared competitive and compatible with existing solar collector concepts are summarized in Figure 5.10 and the results of laboratory feasibility studies of select cycles are reported in Table 5.3. Further descriptions of the select cycles and the cycle-specific material and reactor technologies are discussed in Section 5.3.

5.2.3 High-Temperature Electrolysis

There are basically three main technologies to electrolyze water: (1) alkaline electrolyzers, (2) proton exchange membrane (PEM) electrolyzers,

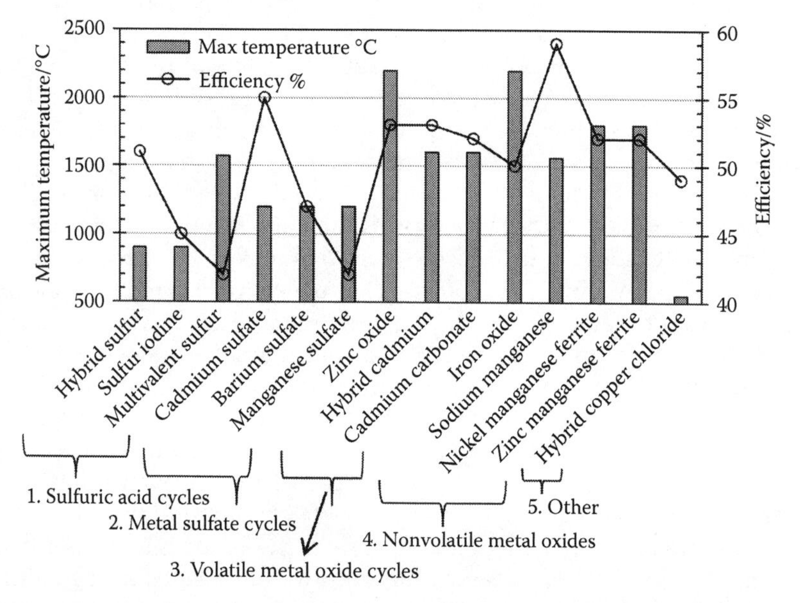

FIGURE 5.10
Thermochemical cycles compatible with existing solar collector concepts deemed competitive after multiple screening. (After Heske, C. et al., UNLV Research Foundation, Las Vegas, NV, 2011.)

TABLE 5.3

Results of Laboratory Feasibility Studies of Select Thermochemical Cycles

Thermochemical Cycles	Results of STCH Study by the DOE
1. Sulfate cycles	Abandoned due to hybrid nature requiring electricity and high temperature sulfate decomposition
2. Cadmium oxide and Zn/ZnO cycles	Study continued under subcontract through Sandia National Laboratories
3. Hybrid CuCl	Studies are continuing at Argonne National Laboratory under sponsorship of the DOE Hydrogen Research Program

Source: Heske, C. et al., UNLV Research Foundation, Las Vegas, NV, 2011.

and (3) high-temperature electrolyzers. One of the reasons for the small-scale industrial production of H_2 via the conventional alkaline water electrolysis is the high electrical energy consumption to the tune of 4.5–5 kWh/m_n^3 H_2 at an electrolysis temperature of ~80°C (Stojić et al. 2003). HTE of steam on the other hand has the potential to reduce the electrical energy requirement significantly, as the positive $\Delta G°(T)$ value for the water decomposition reaction (5.1) goes down linearly with the increase in temperature of the reaction, as is clear from Figure 5.2. The HTE has further advantages of fast reaction kinetics and much cheaper catalyst as compared with the conventional alkaline water electrolysis. For example, a numerical model (Udagawa et al. 2007) demonstrated that at a moderate current density of 700 mA/cm^2 and a cathode inlet stream at a temperature of 1023 K, the electrolysis voltage requirement would be 1.30 V, amounting to an electrical energy consumption rate of ~3.1 kWh per normal m^3 of H_2. Thus, an energy saving of 35% over the low temperature commercial systems is possible.

The most common electrolytes for the HTE are based on oxygen ion conductors such as the yttria-stabilized zirconia (YSZ) and doped ceria, employing metal cermet such as Ni-YSZ as cathode and lanthanum manganite as anode. Thus, the solid oxide electrolysis cell (SOEC) operates in reverse to a solid oxide fuel cell (SOFC). The latter employs the same electrolyte material as well as the same electrode materials for its positive and negative electrodes. The negative electrode (Ni-YSZ) serves as anode in SOFC and as cathode in SOEC and forms the fuel side of the electrode. Although the efficiency of HTE systems is superior to that of the conventional low-temperature electrolysis, the inadequate long durability of oxide-based SOECs is in the way to commercialization of this technologies (Gómez and Hotza 2016; Laguna-Bercero 2012). Proton conducting electrolytes are also being actively investigated for their use in SOEC, as was recently reviewed in Bi et al. (2014) and Malavasi et al. (2010).

5.2.4 Reforming

Reforming of methane, NG, or hydrocarbons is one of the efficient and widely used method for production of hydrogen (Steinfeld and Weimer 2010). Methane reforming is an endothermic reaction in which methane reacts with either steam or CO_2 to produce syngas, mainly comprising of H_2 and CO in the presence of a nickel-based catalyst. The choice of CO_2 reforming or steam reforming is decided by the end product desired. Steam reforming is generally preferable if H_2 production is desired; otherwise, for methanol production, amount of CO_2/steam is varied to get optimal CO/H_2 in the syngas. Conventional methane reforming process involves steam/CO_2 reforming, water gas shift reaction, and separation of H_2/CO_2 mixture (Sheu et al. 2015).

$$CH_4 + H_2O \rightarrow 3H_2 + CO \quad \Delta H° = 206 \text{ kJ/mol} \tag{5.6}$$

$$CH_4 + CO_2 \rightarrow 2H_2 + 2CO \quad \Delta H° = 247 \text{ kJ/mol} \tag{5.7}$$

The endothermic reactions (5.6) and (5.7) represent steam reforming and carbon dioxide reforming, respectively, and are operated within a temperature range of 1073–1273 K.

The CO produced in the above reactions is utilized in water gas shift reaction to produce H_2 as follows:

$$CO + H_2O \rightarrow CO_2 + H_2 \quad \Delta H° = -41 \text{ kJ/mol} \tag{5.8}$$

Reaction (5.8) is slightly exothermic and dual-catalyst beds with interbed cooling are employed for the CO conversion. The first bed operates nearly at 623 K and is packed with high-temperature shift catalyst of Cr-promoted iron oxide. The second bed operates at approximately 523 K and is loaded with Cu-promoted ZnO.

The final step is the separation of the H_2–CO_2 mixture using pressure-swing adsorption (PSA) (Hassan et al. 2008) with molecular sieves or active carbon, which adsorbs all components except H_2. Processes like depressurization of sorbent bed and purging of H_2 are employed for the regeneration of adsorbent. Major challenge associated with PSA technique is that nearly 20% of hydrogen is lost. Other techniques for CO_2 capture and separation include physical and chemical absorption, low-temperature distillation, and gas-separation membranes.

Nowadays, there is an increasing demand for use of solar energy for the production of chemical fuels. Solar methane reforming has been demonstrated on a pilot scale at many places (Spiewak et al. 1993; Sugarmen et al. 2004; Steinfeld et al. 2001; Tamme et al. 2001). Solar reactors are broadly classified into two types: directly and indirectly heated reactors. In SOLASYS project, a pilot plant up to 300 kW was developed, and researchers demonstrated steam reforming of NG in a solar tower facility using porous ceramic foam structures as absorbers coated with Rh as catalyst. SOLASYS is an example of directly heated reactor where reactants are directly exposed to solar radiation. The ASTERIX project utilizes an indirectly heated reactor, where air was used as a heat transfer medium. The maximum methane conversion of 91% approximately at 1073 K was reported by ASTERIX group (Böhmer et al. 1991; Spiewak et al. 1993).

5.2.5 Gasification

The steps involved in a conventional gasification process is similar to those of methane reforming process, which includes an endothermic gasification reaction, followed by water gas shift reaction and separation of H_2/CO_2 mixture depending upon the end uses. The major difference in gasification process is the use of solid materials like coal and biomass for conversion to syngas (H_2 and CO). The coal gasification process is an endothermic reaction requiring energy input of 175 kJ/mol, which proceeds as:

$$C(s) + H_2O(g) \rightarrow CO(g) + H_2(g) \tag{5.9}$$

The second step (water gas shift reaction) is similar to reaction (5.8) of methane reforming to produce syngas, followed by separation of mixture to get H_2.

Solar coal gasification has been performed worldwide with projects like SYNPET. The project SYNPET (Z'Graggen et al. 2006, 2007) demonstrated the development and testing of a 5-kW solar vortex flow reactor. Steam gasification was carried out with materials such as petroleum coke, petroleum coke slurry, and petroleum residues using solar energy.

5.2.6 Cracking

Methane reforming process discussed above has a major drawback of CO_2 formation as the end product, which contributes to greenhouse gas emissions. So, alternative routes for H_2 production must be explored, and more recent studies have considered thermal cracking of NG, oil, and hydrocarbons as an alternative option to steam reforming using concentrated solar energy as the input for the endothermic decomposition reaction.

Thermal cracking of methane has been applied for CO_2-free production of H_2 along with a high-value nanocomposite called carbon black (CB). The thermal decomposition reaction (5.10) is an endothermic process with a standard enthalpy of formation ($\Delta H°$) of 75.6 kJ/mol.

$$CH_4(g) \rightarrow C(s) + 2H_2(g) \tag{5.10}$$

Separation of methane and hydrogen is easier than that of CO_2 and CO (Hufton et al. 2000). The H_2 produced in methane cracking is CO-free, which is a requirement on the gas for use in proton exchange membrane fuel cell in order to avoid CO-poisoning of Pt catalyst. The solid carbonaceous product has various applications such as a reducing agent in metallurgical industry.

Cracking of methane is a more simplified process in comparison to steam reforming. Steps like water gas shift reaction and preferential oxidation of CO in reforming process are absent in cracking. The energy requirement in methane cracking is significantly lower than that in steam reforming process (for steam reforming $\Delta H^\circ_{298.15\,K} = 253.2$ kJ/mol and for methane cracking $\Delta H^\circ_{298.15\,K} = 74.8$ kJ/mol).

Various research work has been conducted by using either a metallic or a carbonaceous catalyst in order to operate the decomposition reaction at lower temperature and improve process kinetics (Avdeeva et al. 1999; Aiello et al. 2000; Figueiredo et al. 2010; Rubbia and Salmieri 2012). Metal catalysts like nickel, cobalt, or iron are deposited on SiO_2 or Al_2O_3 substrates. The combination of metal and support dictate the performance of catalysts to some degree. However, the major challenge associated with the use of catalysts is the regeneration after deactivation due to carbon deposition (Amin et al. 2011).

The catalytic cracking of methane for commercial hydrogen production was first introduced in 1966 and the process was known as HYPRO (Pohlenz and Scott 1966). The process was operated at temperatures of up to 980°C and 1 atm pressure.

Methane cracking is not economically feasible in the current scenario, with limited industrial production in comparison with other well-established methods like methane steam reforming. However, as a source of CO_2-free H_2, methane cracking is becoming increasingly popular among researchers (Amin et al. 2011).

5.3 State of the Art

The noncyclic thermochemical processes to produce hydrogen have been commercialized for long. If the source material is biomass, these processes are renewable. Cyclic thermochemical routes to hydrogen production are the technologies of the future, operating on water as a source material. In this section, the major bottlenecks to the thermochemical processes, namely, the active material and the reactor technology, are discussed.

5.3.1 Materials

Primary roadblocks in realizing the solar thermochemical processes are identified to be the inadequate durability and reactivity of the active materials. The challenges are cycle specific. For example, corrosion of construction materials and the choice of a suitable catalyst material for the high-temperature decomposition of the sulfuric acid at $T > 1000$ K is of serious concern to all of the thermochemical processes in the sulfur group. In metal oxide processes, due to a very high temperature requirement for the thermal decomposition of the higher valent metal oxide at $T \sim 2000$ K, *cycleability* of the active material is found to be inadequate due to sintering and/or onset of melting. The material aspect of the representative oxide processes, sulfur-based cycles, and the Cu/CuCl cycle have been thoroughly discussed in Roeb et al. (2012). In this section, a short update on the material-related challenges in the most promising cycles such as ferrite, Zn–ZnO, sulfur-based cycles are presented.

5.3.1.1 Ferrites

In the late 1970s, an iron oxide-based redox cycle was first proposed by Nakamura (1977). This is one of the earliest and widely used materials for research on solar hydrogen generation. The two-step cycle consisting of thermal reduction and a subsequent hydrolysis of the reduced product proceeds as follows:

$$Fe_3O_4 \rightarrow 3FeO + 0.5\ O_2 \tag{5.11}$$

$$3FeO + H_2O \rightarrow Fe_3O_4 + H_2 \tag{5.12}$$

From the thermodynamics point of view, high-temperature thermal reduction (around 2573 K) of magnetite is accompanied by an enthalpy change of ΔH°_{298K} = 319.5 kJ/mol and the hydrolysis (around 900 K) of wustite is mildly exothermic ($\Delta H_{298K} = -33.6$ kJ/mol) (Kodama and Gokon 2007). However, melting point of FeO (1643 K) is lower than the temperature of thermal reduction, resulting in the formation of liquid FeO phase. In practice, FeO will melt and solidify alternately in each cycle. To address this issue, mixed solid solutions of redox pair Fe_3O_4/FeO have been studied with respect to their H_2 generation capability. A divalent transition metal M (M = Mn, Co, Ni, or Zn) partially substitutes the Fe^{2+} lattice site in Fe_3O_4. After substitution, the ferrite $\{(Fe_{1-x}M_x)_3O_4\}$ forms $\{(Fe_{1-x}M_x)_{1-y}O\}$ on reduction, which undergoes the hydrolysis reaction to yield H_2. In 2007, Kodama et al. (2008) reported coating *m*-ZrO_2 (monoclinic zirconia) with nickel-based ferrites to avoid the problem of sintering at the high-temperature reduction process. The *m*-ZrO_2 particles were coated with $Ni_xFe_{3-x}O_4$ ($x = 1$, 0.35, and 0.65) and the cyclic reactions of high-temperature reduction and hydrolysis were repeatable at 1673 and 1273 K, respectively. Previous studies on unsupported Ni and Co ferrites indicated the formation of dense, nonporous, and hard particles after thermal reduction step (Kodama et al. 2002, 2005). These particles were difficult to pulverize and subsequent water splitting step could not be performed. However, in the case of supported ferrites, pulverization and the following hydrogen generation step could be performed due to the formation of soft, brittle, and fine porous particles. Supported ferrites (*m*-ZrO_2) effectively addressed the issue of severe sintering of particles during the thermal reduction process. $NiFe_2O_4$ was found to be the most suitable among the tested ferrites, capable of high hydrogen yield over several runs. From XRD results (Kodama et al. 2008), it was concluded that the reduced phase of the unsupported Ni(II) ferrite was less reactive than that of $NiFe_2O_4$/*m*-ZrO_2 in the water splitting step, as the peak of wustite phase disappeared after water splitting in case of supported ferrites. This is because of the severe sintering of the unsupported ferrite particles. Figure 5.11 depicts comparison between amounts of H_2 generated in different ferrite systems. The authors (Kodama et al. 2008) compared the reactivity of $NiFe_2O_4$/*m*-ZrO_2 with other metal-doped ferrite/*m*-ZrO_2 systems such as Mn-ferrites, Mg-ferrites, and Co–Mn-ferrites. Among all these systems, $NiFe_2O_4$/*m*-ZrO_2 was found to have the greatest reactivity. The BET surface area of $NiFe_2O_4$/*m*-ZrO_2 system reduced significantly to 0.4 m^2/g after the first cycle run from the initial value of 13.8 m^2/g. However, no further deterioration in the surface area was noted up to the completion of the sixth cycle. The sample was prevented from severe sintering, and pulverization can be done by using a pestle and mortar.

Kodama and colleagues (Gokon et al. 2007; Kodama et al. 2006, 2009) studied YSZ support for FeO–Fe_3O_4 system and found a promising redox material

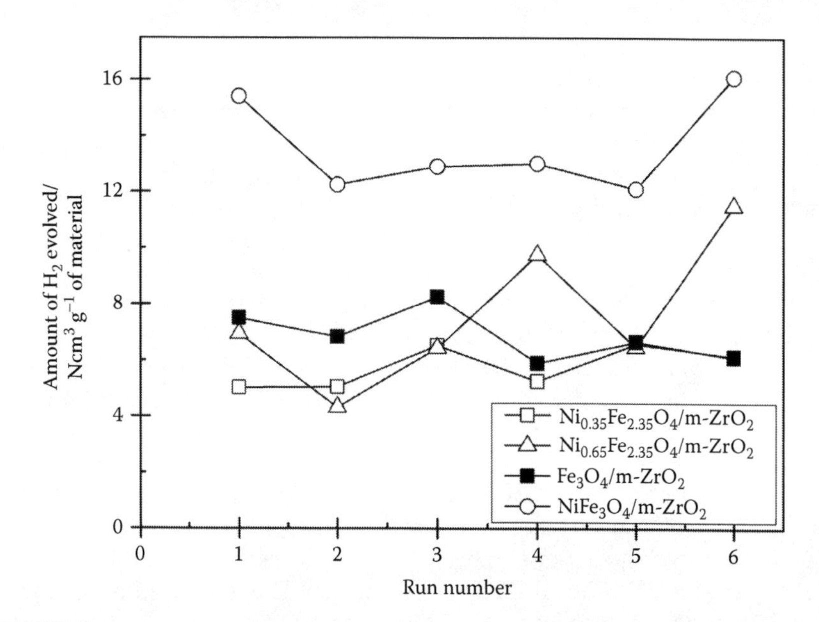

FIGURE 5.11
A comparision between amount of H_2 evolved in different ferrite systems. (Reprinted with from *Solar Energy*, 82, Kodama, T., N Gokon, and R Yamamoto, Thermochemical Two-Step Water Splitting by ZrO2-Supported NixFe3–xO4 for Solar Hydrogen Production, 73–79. Copyright [2008] , with permission from Elsevier.)

of Fe-containing YSZ for the thermochemical cycle. Different redox materials and cyclic reactions were reported depending on the amount of doping of Y_2O_3. For Fe_3O_4 supported on YSZ (doped with 3 mol% of Y_2O_3), during the reduction step at 1673 K under an inert atmosphere, Fe^{2+} enters into the stabilized cubic ZrO_2 crystal to form Fe^{2+}-containing YSZ (Fe^{2+}-YSZ). In the subsequent reaction with water, formation of Fe_3O_4 crystals was observed. XRD studies on the Fe_3O_4/YSZ have confirmed this reaction mechanism. When YSZ doped with more than 8 mol% of Y_2O_3 was used as the support for Fe_3O_4, the reaction mechanism of the hydrolysis changed. In the very first high-temperature reduction process, the Fe^{2+} ions in the YSZ lattice were converted to Fe^{3+} ions in the hydrogen production step. This change in mechanism of oxidation of Fe^{2+} ions into Fe^{3+} ions occurs due to the increased concentration of Y ions in zirconia, stabilizing the Fe^{3+} ions in the cubic crystal structure. It was found that when doping of Y_2O_3 is increased (>8 mol%), it resulted in more stable H_2 production in comparison to ZrO_2-supported ferrites. The YSZ-supported ferrites do not change the crystal structure because of temperature fluctuations, whereas nonstabilized ZrO_2 may transform between monoclinic and tetragonal structure.

Ehrensberger et al. (1995) studied the effect of manganese in Fe_3O_4/FeO system. $(Fe_{1-x}Mn_x)_{1-y}O$ (x = 0.0, 0.1, and 0.3) was tested under above-prescribed conditions and oxidized to $(Fe_{1-x'}Mn_{x'})_3O_4$ with $x' < x$, where x' is the stoichiometry factor determined from the literature data assuming that lattice parameter obeys Vegard's law. It was observed that partial iron substitution with 10 and 30% (on molar basis) manganese in the wustite phase changed the reduction kinetics; however, it did not affect the rate of hydrogen evolution in hydrolysis reaction. The reaction rates for the different oxides decreased with increase in Mn content. If manganese content is increased, the number of defects is lowered. As the number of defect sites determine the diffusion rate, the reactivity of magnowustite $\{(Fe_{1-x}Mn_x)_{1-y}O\}$ decreases with high manganese content.

5.3.1.2 Zinc/Zinc Oxide Cycle

A two-step cycle based on Zn/ZnO is one of the most promising thermochemical cycle due to large change in entropy during the thermal reduction step. The thermal reduction (2173 K) and the subsequent hydrolysis step (773 K) are represented by Equations (5.13) and (5.14), respectively:

$$ZnO(s) \rightarrow Zn(g) + 0.5O_2(g) \tag{5.13}$$

$$Zn(s) + H_2O \rightarrow ZnO(s) + H_2(g) \tag{5.14}$$

Reaction (5.13) is endothermic, requiring a pure thermal energy input of 450 kJ/mol at 2173 K.

The reaction is thermodynamically favored at lower temperature if pressure is decreased or carrier gas such as argon is introduced. Reaction (5.14) is mildly exothermic in nature, releasing 130 kJ/mol of energy. The exergy efficiency up to 82% is obtained, assuming complete heat recovery between the high- and low-temperature process steps (Perkins et al. 2008).

The melting point and boiling point of zinc is 692 and 1180 K, respectively. So, after thermal reduction, there is a serious problem of recombination of gas stream [Zn(g) and O_2] at the temperature greater than 1180 K. To address this issue, the product stream must be rapidly quenched in a quenching apparatus. Gstoehl et al. (2008) proposed a quenching apparatus to avoid recombination of product gas stream and maximum Zn yield of 94% was obtained for Ar/Zn(g) dilution of 530 at a cooling rate of 10^5 K/s. The details of quenching apparatus is discussed in Section 5.3.2.

The efficiency of oxidation (hydrolysis) step is limited due to the formation of ZnO on the surface of metal (Perret 2011). This prevents further oxidation of zinc metal, thus decreasing the efficiency of the step. Another important criterion is the size of Zn particles, as efficiency depends on the ratio of particle surface area to volume of particles. Submicrometer Zn particles are desired for this step, which are produced from rapid quenching step. However, efficient recovery after quenching step is a challenge for researchers.

Other issues associated with Zn/ZnO thermochemical cycle are materials for construction of reactor that can withstand high temperature reduction and deposition of Zn particles on reactor walls.

5.3.1.3 Sulfur-Based Cycles

Sulfur-based families consists of two extensively studied cycles—sulfur iodine (S-I) and hybrid sulfur (HyS) (Banerjee et al. 2007; Corgnale and Summers 2011; Dehghani and Sayyaadi 2013; Kubo et al. 2004; Roeb et al. 2012; Sub et al. 2012; Summers et al. 2008; Takai et al. 2011). Both of these water splitting cycles for H_2 generation deals with corrosive environment at elevated temperatures. The S-I cycle consists of three steps:

1. Production of HI and H_2SO_4,

$$2H_2O + SO_2 + I_2 \rightarrow H_2SO_4 + 2HI \quad (\text{Bunsen reaction at } 300-400\text{ K}) \tag{5.15}$$

2. Decomposition of HI at 400 – 1000 K,

$$2HI \rightarrow H_2 + I_2 \tag{5.16}$$

3. The decomposition of sulfuric acid in a two-step process as below:

$$H_2SO_4(aq) \rightarrow H_2SO_4(g) \rightarrow SO_3(g) + H_2O(g) \quad (623-673\text{ K}) \tag{5.17}$$

$$SO_3(g) \rightarrow SO_2(g) + 0.5\,O_2(g) \quad (973 - 1273\ K) \tag{5.18}$$

The product of Bunsen reaction (5.15) is a mixture of HI and unreacted $H_2O(g)$ and I_2. The HI separation from the product mixture is difficult and is performed by techniques like reactive or extractive distillation. Extractive distillation is a multistep process and done by using phosphoric acid, whereas reactive distillation is a single-step process wherein both separation and decomposition of HI occur in a single reaction vessel at 583 K and 51 atm (Wong et al. 2007). HI is a highly corrosive acid, and its effect on reaction chambers, heat exchangers, and other key components of HI decomposition at higher temperature is severe. So, choice of safe and durable construction materials for HI decomposition is a key challenge for scientists.

Wong et al. (2007) reported about the construction materials suitable for highly corrosive environment of HI coupled with high temperature. Four classes of corrosion-resistant materials—refractory materials, ceramics, superalloys, and reactive metals—were studied for immersion coupon corrosion test for their suitability under corrosive environment applications. Refractory materials such as Ta, Ta-40 Nb, and Nb-7.5 Ta were found to tolerate extreme corrosive environment. Optical micrographs of all the tested refractory coupons revealed the presence of a uniform passive oxide layer on the surface. The growth of these oxide layers was found to be directly related to immersion time and test temperature. Very fine pits at polishing grooves were observed in Nb-7.5 Ta coupons at boiler condition. However, presence of these pits resulting from stress corrosion cracking caused by polishing does not affect material behavior to a greater extent. No observable changes were reported for Ta and Ta-40 Nb coupons other than passivation. SEM micrographs of Ta and Ta-40 Nb confirm stable surfaces after immersion tests. No weight loss was observed for all the tested refractory coupons, signifying no metal dissolution under harsh corrosive conditions. However, extensive pitting and metal dissolution were observed along with a nonuniform passive layer for reactive metals like Zr702 and Zr705, thus making these materials unsuitable for corrosive HI environment. Significant weight loss of 31% was observed for C-276 test coupon (a nickel-based super alloy), depicting an extremely poor corrosion resistance of Ni-based superalloys under corrosive HI environment. No pitting was observed for mullite (a ceramic-based material), making it viable for containment materials for interconnect components such as piping and tubing.

Unlike the S-I cycle, hybrid sulfur cycle (also known as Westinghouse cycle) consists of two steps: (1) decomposition of sulfuric acid as represented by reactions (5.17) and (5.18) and (2) electrolysis of sulfur dioxide and water at 0.17 V and at temperature of 353 K as per the overall electrochemical reaction (5.19):

$$SO_2 + 2H_2O \rightarrow H_2 + H_2SO_4 \tag{5.19}$$

The decomposition of sulfuric acid step is common in both HyS and S-I cycles. Similar to HI, sulfuric acid, being corrosive in nature, possesses extreme material challenges for the construction of key components such as the H_2SO_4 evaporator and SO_3 decomposer. In HycycleS (a European-funded project), researchers reported SiC and siliconized SiC (SiSiC) suitable as construction materials for the decomposition of sulfuric acid after performing corrosion tests on the specific materials for more than 5000 h under boiling sulfuric acid environment (Roeb et al. 2013). Various scalable prototypes with construction material for decomposer such as SiC were developed and successfully tested, emphasizing the durability of these materials. Within the HycycleS project, investigations are ongoing about the use of YSZ as material for oxygen membranes for the removal of O_2 in order to shift the equilibrium toward increased decomposition (according to Le Chatelier's principle) as per reaction (5.18). However, the performance and stability of YSZ-based membranes for O_2 removal were not found suitable for practical applications.

Another important issue relevant to sulfur-based cycles is the materials for catalysts. The decomposition reaction of sulfur trioxide to sulfur dioxide occurs at 1200 K under the severe corrosive environment. To increase the reaction rate for this endothermic reaction under those harsh operating conditions, the choice of suitable catalyst material is an absolute necessity. Studies under HycycleS project revealed that a Fe–Cr mixed alloy ($Fe_{0.7}Cr_{1.3}O_3$) was most promising among all the materials examined. Giaconia et al. (2011) evaluated catalytic activity of siliconized silicon carbide honeycombs coated with Fe_2O_3 during SO_3 decomposition reaction. The investigation was conducted using a laboratory-scale apparatus operated at 1123 K and 1 atm pressure for a period of 100 h. The results revealed higher SO_3 conversion of around 80% and insignificant deactivation of the material over test period of 100 h, thus making SiSiC-coated Fe_2O_3 a promising catalyst candidate. Investigations on the catalytic decomposition of HI indicated Pt (Shindo et al. 1984) and activated carbon (Ginosar et al. 2011) as suitable candidates. However, due to high cost and limited availability, Pt is not suitable for large-scale H_2 production plants. Fu et al. (2016) studied the effect of source of raw material on the catalytic activity of activated carbon for HI decomposition. In this study, five activated carbon samples generated from wood, coconut shell, shell, bamboo, and coal were probed using techniques like XRD, SEM, BET, proximate analysis, and ultimate analysis. It was found that activated carbon produced from coconut shell and shell showed highest activity.

5.3.2 Reactor Technology

Solar thermochemical reactors are heat engines that produce fuels such as hydrogen or CO from concentrated solar energy, water, or CO_2. The efficiency of solar reactors plays a key role in solar fuel generation. Subsequent production of liquid synthetic fuels from H_2 and CO gas mixture (syngas) was deemed to be a viable replacement to petroleum derivatives used in transportation by utilizing less than 0.7% of the U.S. land area (Ermanoski et al. 2013). It is noteworthy that a mature biofuel-based alternative would require a land area of over 100 times to match the energy output from the solar thermochemical reactors. In the past few years, researchers have demonstrated numerous new designs and concepts and analyzed key performance parameters for improving the output of solar reactors (Davidson 2009; Djamal and Chakib 2015; Ermanoski et al. 2013; Gstoehl et al. 2008; Konstandopoulos and Agrofiotis 2006; Meier and Steinfeld 2008; Perkins et al. 2008). In the following, numerous studies conducted in the development of solar reactors for H_2 production in the last decade are discussed.

Ermanoski et al. (2013) demonstrated the use of a novel packed bed reactor to produce fuels such as CO or H_2 using CeO_2 water splitting thermochemical cycle with an efficiency in excess of 30%. The reactor consisted of three chambers: a thermal reduction chamber, a recuperator, and a fuel production unit, as depicted in Figure 5.12. The oxidized particles were thermally reduced by using concentrated solar energy entering through the aperture, producing oxygen, which was continuously removed using a pump. The reduced

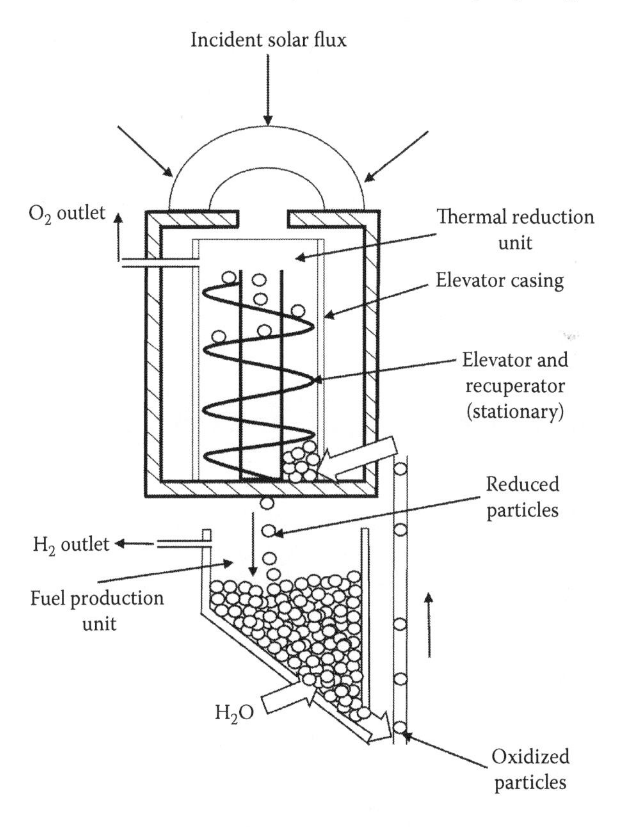

FIGURE 5.12
Schematic drawing of a moving packed bed reactor. (After Ermanoski, I., N. P. Siegel, and E. B. Stechel, *Journal of Solar Energy Engineering*, 135, 031002, 2013.)

particles moved continuously downward into the fuel production unit via a connecting recuperator tube. These particles were exposed to H_2O or CO_2 in the fuel production unit, and the oxidized particles were brought to the base of the recuperator using a return elevator. These re-oxidized particles were then elevated toward the inlet of reduction chamber using a return elevator. These particles moved toward the thermal reduction chamber in a counter flow arrangement with regard to reduced particles in order to enable the heat transfer via conduction mechanism, which helps in achieving maximum possible heat recovery between the high temperature reduction and low temperature fuel production chambers. The sensible heat recovery by using counter current heat transfer mechanism between reduced and oxidized particles helps in improving the efficiency of the reactor. This reactor also implemented unique pressure separation mechanism by having separate chambers for the reduction and fuel production with a seal provided by the packed bed. Pressure separation between the two chambers prevented products from two cycle steps from mixing, eliminating recombination and some separation losses. In addition to this, pressure separation plays a key role in use of vacuum pumping to reduce the partial pressure of oxygen in thermal reduction chamber. Existing reactor concepts use inert gas sweeping to decrease the oxygen partial pressure in the reduction chamber (Chueh et al. 2010; Diver et al. 2008; Kodama et al. 2008), but studies show that new methodology of vacuum sweeping helps in improving performance of the reactor.

Agrafiotis et al. (2005) first reported the use of monolithic honeycomb reactors for iron oxide-based system. Ceramic (SiC) honeycomb structures having dimensions of $\phi 144 \times 200$ mm were coated with active redox agent (ferrites). A wide variety of ferrites were synthesized and comparatively evaluated for their water splitting and regeneration capability in a two-step cycle, where thermal reduction and hydrolysis reaction occurred, respectively, at 1473 and 1073 K.

The research group at German Aerospace center, DLR Germany, developed a solar reactor for mixed iron oxides (Roeb et al. 2006). The reactor, which is made up of a stainless steel housing, can withstand a temperature of up to 1673 K. The reactor had a quartz window for letting concentrated solar radiation to come in and an exhaust for letting the gases to exit. The researchers at DLR successfully tested repeatability and H_2 generation of iron-based systems up to 10 cycles using this reactor. An economic evaluation of reactor was undertaken considering various operating constraints, and the production cost of H_2 was projected to be nearly 0.18 €/kWh.

HYDROSOL projects are European Union-funded projects with an objective of producing renewable hydrogen using concentrated solar energy with a specific thermochemical cycle. Till now, three HYDROSOL projects (I, II, and 3D) have been proposed to address the issues associated with fossil fuel-based hydrogen production technologies and to harness huge potential of conversion of solar radiation to chemical fuels. A honeycomb monolith solar reactor coated with water splitting materials was developed under HYDROSOL-I project. Konstandopoulos and Agrafiotis addressed the issue of intermittent H_2 production, with no yield when the material is being regenerated during the HYDROSOL-I project. In the HYDROSOL-II project, an innovative pilot-scale solar reactor capable of producing H_2 continuously for 2 days was developed and achieved a maximum repeatability up to 40 cycles (Konstandopoulos and Agrofiotis 2006). Reaction efficiency up to 28% and a process efficiency up to 9% were reported for this pilot-scale solar plant. Konstandopoulos and Agrafiotis have proposed a novel reactor concept called "conti-reactor." The conti-reactor is a dual-chamber reactor, where the water splitting reaction occurs in one chamber and regeneration occurs in the other, thus enabling a continuous production of solar hydrogen. These reactors produce H_2 in alternate cycles, and it was so successful that HYDROSOL-3D (demonstration of 1 MW solar plant) was implemented in 2011 (Roeb et al. 2011).

To address the issue of Zn(g) and O_2(g) recombination in ZnO–Zn volatile cycle, a quenching apparatus based on the concept of three-temperature regions has been proposed by Gstoehl et al. (2008), as shown in Figure 5.13. The quench apparatus consists of three regions: (1) an inlet hot zone at a temperature above the ZnO decomposition temperature to suppress the formation of ZnO; (2) a transition zone at the temperature above the Zn saturation but below the decomposition of ZnO (in this zone, inert gas such as Ar is passed, thus avoiding Zn(g)/O_2 reaction at the walls); and (3) an outlet cold zone, where temperature is below Zn saturation due to injection of cold Ar gas and water cooling of the wall. This cold outlet region assisted in homogeneous nucleation of Zn(g), which was prone to oxidation. Experiment was performed with ZnO sample kept inside the cavity of a solar reactor, where dissociation of ZnO was monitored by online thermogravimeter. The quenching apparatus was connected to the end of solar reactor where the product

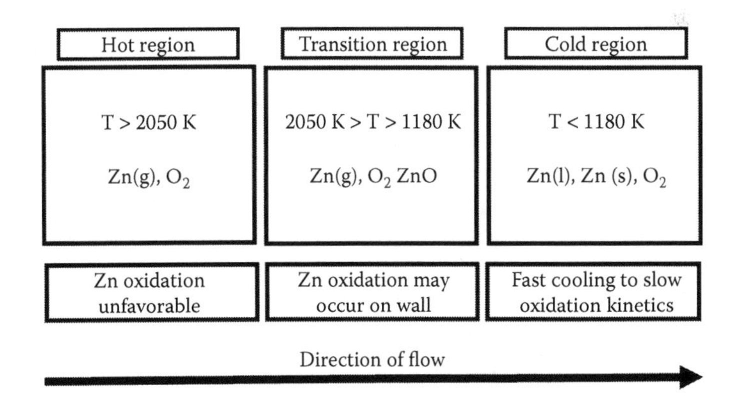

FIGURE 5.13
A schematic of quench concept based on three temperature regions. (After Gstoehl, D. et al., *Journal of Materials Science*, 43, 4729–4736, 2008.)

gases were quenched onto water-cooled surfaces at cooling rates from 20,000 to 120,000 K/s, aided by injection of a cold stream of Ar gas, thereby suppressing the formation of the metal oxide. Results confirmed that zinc yields depend on Zn(g) dilution by Ar and the surface temperature where products are quenched. It was observed that Zn yields in the range of 40–94% were obtained for Ar/Zn(g) dilutions of 170–1500. Maximum Zn yield of 94% was obtained for Ar/Zn(g) dilution of 530 at a cooling rate of 10^5 K/s.

A highly efficient solar reactor known as STARS (solar thermochemical advanced reactor system) was successfully developed and tested by a team of researchers at Pacific Northwest National Laboratory (PNNL) under a U.S. DOE project (Zhenga et al. 2015). STARS (a technical readiness level 4 prototype) consisted of solar methane reforming reactor made up of Haynes 214 (an alloy of nickel, chromium, iron, and aluminum) and heat exchangers, which were mounted inside a nacelle near the dish focus, with thermocouples embedded inside the reactant wall 6.34 mm away from the heated surface. Concentrated solar power passed through the reactor wall and heated the reaction mixture inside the reactor microchannels. The combination of meso- and microchannels technologies reduces heat and mass transfer resistance inside the reactor channels, resulting in high heat flux, low thermal losses, and high receiver efficiencies. Researchers at PNNL demonstrated solar chemical energy conversion efficiency (ratio of increase in higher heating value in product stream to the incident solar flux on parabolic dish concentrator) as high as 69% with steam methane reforming process. Reactor temperature greater than 1073 K and high direct normal insolation (DNI > 860 W/m^2) were found to be favorable for high solar chemical energy conversion efficiency. However, steam-to-carbon ratio do not influence the conversion efficiency. It was also observed that the output gas stream (H_2 + CO) of the reactor has about 25–28% more chemical energy than the incoming methane feed. With the commercialization of this reactor technology, electricity cost less than 6 €/kWh can be achieved by 2020.

5.4 Conclusion

With continued, ever-increasing rate of depletion of the finite fossil fuel resources, security and safety-related concerns in nuclear fission reactor, and the uncertainty in the fruition of global effort on fusion, finding an efficient way to harness the infinite and intermittent solar insolation is the only reliable and environment-friendly means to ensure the energy security in the future. Due to intermittency of solar radiation, the energy needs to be stored in a chemical fuel such as hydrogen. The primary method of commercial hydrogen production is through hydrocarbon reforming, which requires a large amount of heat input. The heat requirement of the reforming process is often met by burning a portion of the fuel being reformed. Solar thermochemical routes to hydrogen production meet their heat demand by concentrated solar radiation, and these technologies are still under development. The *noncyclic* routes to hydrogen production such as reforming, gasification, and cracking of a carbonaceous fuel are well established. With the emergence of efficient solar thermal collection technologies, solar thermal power generation has achieved an efficiency of approximately 25%. If the solar electric power is used in an 80% efficient alkaline water electrolysis to produce hydrogen, a solar-to-H_2 energy conversion efficiency of 20% is reachable by the existing technologies. The challenge now is to match or better exceed this efficiency mark (i.e., 20%) in solar-to-storable hydrogen conversion.

It is widely believed in the research community that *cyclic* thermochemical routes to hydrogen production from water can attain a solar-to-hydrogen conversion efficiency in excess of 30%. The major roadblocks in commercializing the thermochemical cycles are materials (active as well as structural) and reactor technologies, both being highly interlinked and specific to the chosen cycle. Broadly speaking, there are multistep, low-temperature thermochemical cycles that have to address the problems of (1) energy leakage due to a large change in temperature at each step and (2) H_2 separation from a mixture under cycle-specific harsh operating conditions. In such cycles, suitable materials for reactor in contact with corrosive active material, an appropriate membrane for gas separation, and an optimum reactor design are being actively investigated.

The most promising of thermochemical cycles are the two-step, high-temperature cycle operating the thermal reduction step at nearly 2000 K. Due to high temperature and the fact that water-to-H_2 conversion process is subject to Carnot limitation, the *thermal* efficiency in such cycles is high. However, the *collection* efficiency of the optical system goes down with temperature. Nonetheless, the overall *system* efficiency, which is a product of the collection and thermal efficiency, is higher as compared with a low-temperature process. Moreover, nonvolatile cycles such as the ferrite- and ceria-based cycles produce H_2 and O_2 at the thermal oxidation and reduction steps, respectively, at different temperatures. Thus, the product gas H_2 is obtained in pure form, obviating any need for an energy-intensive separation process. However, the search for a suitable active redox material is still on, as this has to satisfy a host of conflicting requirements. For example: (1) The active material should have a high diffusion coefficient for faster reaction, but this also leads to the problem of unwanted sintering of the powder and a reduction in interface area. (2) High oxidation/reduction temperature leads to better reaction kinetics, but it is also accompanied with more material losses due to partial volatilization and material-handling issues.

Finally, practicably attainable system efficiency is dictated by both the thermodynamic and kinetics of the active material as well as a reactor with a high recuperation ability. This factor has been beautifully quantified as *utilization* factor (Siegel et al. 2013). Thus, the material and reactor aspects of a process are intimately linked, and both of these together dictate the system efficiency. It is projected that the conventional thermochemical processes such as cracking and reforming will be the first to adopt a solar-based reactor. In the medium to long term, some of the thermochemical cycles will make their way to produce the hydrogen from all renewable sources like water and biomass using solar insolation for heat and, if required, for the electrical energy in a hybrid thermochemical process.

References

Abbasi, T. and S. A. Abbasi. 2011. "'Renewable' hydrogen: Prospects and Challenges." *Renewable and Sustainable Energy Reviews* 15(6): 3034–40.

Agrafiotis, C., M. Roeb, A. G. Konstandopoulos, L. Nalbandian, V. T. Zaspalis, C. Sattler, P. Stobbe, and A. M. Steele. 2005. "Solar Water Splitting for Hydrogen Production with Monolithic Reactors." *Solar Energy* 79(4): 409–21. doi:http://dx.doi.org/10.1016/j.solener.2005.02.026.

Aiello, R., J. E. Fiscus, H. Loye, and M. D. Amiridis. 2000. "Hydrogen Production via the Direct Cracking of Methane over Ni / SiO 2 : Catalyst Deactivation and Regeneration." *Applied Catalysis A: General* 192: 227–34.

Amin, A. M., E. Croiset, and W. Epling. 2011. "Review of Methane Catalytic Cracking for Hydrogen Production." *International Journal of Hydrogen Energy* 36(4): 2904–35. doi:10.1016/j.ijhydene.2010.11.035.

Anon. 2015. "Hydrogen Production Pathways." http://energy.gov/eere/fuelcells/hydrogen-production-pathways.

Avdeeva, L.B., D.I. Kochubey, and Sh.K. Shaikhutdinov. 1999. "Cobalt Catalysts of Methane Decomposition: Accumulation of the Filamentous Carbon." *Applied Catalysis A: General* 177(1): 43–51. doi:10.1016/S0926-860X(98)00250-6.

Banerjee, A. M., K. Bhattacharyya, M. R. Pai, A. K. Tripathi, V. S. Kamble, S. R. Bharadwaj, and S. K. Kulshreshtha. 2007. "Studies on Sulfur-Iodine Thermochemical Cycle for Hydrogen Generation." *BARC Newsletter* no. 285: 67–72.

Baykara, S. Z. 2004. "Experimental Solar Water Thermolysis." *International Journal of Hydrogen Energy* 29(14): 1459–69. doi:http://dx.doi.org/10.1016/j.ijhydene.2004.02.011.

Bi, L., S. Boulfrad, and E. Traversa. 2014. "Steam Electrolysis by Solid Oxide Electrolysis Cells (SOECs) with Proton-Conducting Oxides." *Chemical Society Reviews* 43(24): 8255–70.

Böhmer, M., U. Langnickel, and M. Sanchez. 1991. "Solar Steam Reforming of Methane." *Solar Energy Materials* 24 (1–4): 441–48. doi:10.1016/0165-1633(91)90081-U.

Chase, M. W. 1998. "NIST—JANAF Thermochemical Tables." *Journal of Physical and Chemical Reference Data Monograph*, no. 9: 1963pp.

Chueh, W. C., M. Abbott, D. Scipio, and S. M. Haile. 2010. "High-Flux Solar-Driven Thermochemical Dissociation of CO_2 and H_2O Using Ceria Redox Reactions." *Science* 63: 2010.

Corgnale, C., and W. A. Summers. 2011. "Solar Hydrogen Production by the Hybrid Sulfur Process." *International Journal of Hydrogen Energy* 36(18): 11604–19. doi:http://dx.doi.org/10.1016/j.ijhydene.2011.05.173.

Davidson, J. H. 2009. "Study of a Quench Device for the Synthesis and Hydrolysis of Zn Nanoparticles : Modeling and Experiments." 131: 1–9. doi:10.1115/1.3142825.

Dehghani, S., and H. Sayyaadi. 2013. "Energy and Exergetic Evaluations of Bunsen Section of the Sulfur E Iodine Thermochemical Hydrogen Production Plant."*International Journal of Hydrogen Energy* 38(22): 9074–9084.

Diver, R. B., J. E. Miller, M. D. Allendorf, N. P. Siegel, and R. E. Hogan. 2008. "Solar Thermochemical Water-Splitting Ferrite-Cycle Heat Engines." *Journal of Solar Energy Engineering* 130(4): 041001. doi:10.1115/1.2969781.

Djamal, D., and S. Chakib. 2015. "Review of Recent Advanced in Solar Thermo-Chemical Reactors for Hydrogen Production from Water." *Proceedings of Engineering and Technology* 3–6.

Dresselhaus, M., G. Crabtree, and M. Buchanan. 2003. "Basic Research Needs for the Hydrogen Economy." Technical report, Argonne National Laboratory, Basic Energy Sciences, US DOE, http://www. sc. doe. gov/bes/hydrogen. pdf.

Ehrensberger, K. 1995. "Comparative Experimental Investigations of the Water-Splitting Reaction with Iron Oxide Fe1-yO and Iron Manganese Oxides (Fe1-X Mnx)1-Y O." *Solid State Ionics* 78: 151–60.

Ermanoski, I., N. P. Siegel, and E. B. Stechel. 2013. "A New Reactor Concept for Efficient Solar-Thermochemical Fuel Production." *Journal of Solar Energy Engineering* 135(3): 031002. doi:10.1115/1.4023356.

Figueiredo, J. L., J. J. M. Órfão, and A. F. Cunha. 2010. "Hydrogen Production via Methane Decomposition on Raney-Type Catalysts." *International Journal of Hydrogen Energy* 35(18): 9795–9800. doi:10.1016/j.ijhydene.2009.12.071.

Fletcher, E. A. 2001. "Solarthermal Processing: A Review." *Journal of Solar Energy Engineering* 123(2): 63–74.

Fletcher, E. A., and R. L. Moen. 1977. "Hydrogen-and Oxygen from Water." *Science* 197(4308): 1050–56.

Fu, G., Z. Wang, Y. Zhang, Z. Huang, J. Liu, J. Zhou, and K. Cen. 2016. "Effect of Raw Material Sources on Activated Carbon Catalytic Activity for HI Decomposition in the Sulfur-Iodine Thermochemical Cycle for Hydrogen Production." *International Journal of Hydrogen Energy* 41(19): 7854–7860.

Funk, J. E. 2001. "Thermochemical Hydrogen Production: Past and Present." *International Journal of Hydrogen Energy* 26(3): 185–90. doi:http://dx.doi.org/10.1016/S0360-3199(00)00062-8.

Giaconia, A., S. Sau, C. Felici, P. Tarquini, G. Karagiannakis, C. Pagkoura, C. Agrafiotis, et al. 2011. "Hydrogen Production via Sulfur-Based Thermochemical Cycles: Part 2: Performance Evaluation of Fe_2O_3-Based Catalysts for the Sulfuric Acid Decomposition Step." *International Journal of Hydrogen Energy* 36(11): 6496–509. doi:http://dx.doi.org/10.1016/j.ijhydene.2011.02.137.

Ginosar, D. M., L. M. Petkovic, and K. C. Burch. 2011. "Commercial Activated Carbon for the Catalytic Production of Hydrogen via the sulfur–Iodine Thermochemical Water Splitting Cycle." *International Journal of Hydrogen Energy* 36(15): 8908–14. doi:http://dx.doi.org/10.1016/j.ijhydene.2011.04.164.

Gokon, N., T. Mizuno, Y. Nakamuro, and T. Kodama. 2007. "Iron-Containing Yttria-Stabilized Zirconia System For Two-Step Thermochemical Water Splitting." *Journal of Solar Energy Engineering* 130(1): 11018.

Gómez, S. Y., and D. Hotza. 2016. "Current Developments in Reversible Solid Oxide Fuel Cells." *Renewable and Sustainable Energy Reviews* 61: 155–74. doi:http://dx.doi.org/10.1016/j.rser.2016.03.005.

Gstoehl, D., A. Brambilla, L. O. Schunk, and A. Steinfeld. 2008. "A Quenching Apparatus for the Gaseous Products of the Solar Thermal Dissociation of ZnO." *Journal of Materials Science* 43(14): 4729–36. doi:10.1007/s10853-007-2351-x.

Hassan, M. M., N. S. Raghavan, D. M. Ruthven, and H. A. Boniface. 2008. "Pressure Swing Adsorption." *AIChE journal* 31(12): 2008–16.

Haxel, G. B., J. B. Hedrick, and G. J. Orris. 2002. "Rare Earth Elements—Critical Resources for High Technology." http://pubs.usgs.gov/fs/2002/fs087-02/.

Heske, C., S. Moujaes, A. Weimer, B. Wong, N. Siegal, E. McFarland, E. Miller, M. Lewis, C. Bingham, and K. Roth. 2011. "High Efficiency Generation of Hydrogen Fuels Using Solar Thermochemical Splitting of Water." UNLV Research Foundation, Las Vegas, Nevada. http://www.osti.gov/scitech/biblio/1025597.

Hufton, J., W. Waldron, S. Weigel, M. Rao, S. Nataraj, and S. Sirca. 2000. "Proceedings of the 2000 Hydrogen Program Review NREL/CP-570-28890." *Proceedings of the 2000 Hydrogen Program Review*, 1–14. doi:NREL/CP-570-28890.

International Energy Outlook 2013. 2013. *Report Number: DOE/EIA-0484(2013)*. Citeseer. http://www.eia.gov/forecasts/ieo/index.cfm.

Jellinek, H. H. G., and H. Kachi. 1984. "The Catalytic Thermal Decomposition of Water and the Production of Hydrogen." *International Journal of Hydrogen Energy* 9(8): 677–88.

Kappauf, T., and E. A. Fletcher. 1989. "Hydrogen and Sulfur from Hydrogen sulfide—VI. Solar Thermolysis." *Energy* 14(8): 443–49. doi:http://dx.doi.org/10.1016/0360-5442(89)90111-4.

Kodama, T., and N. Gokon. 2007. "Thermochemical Cycles for High-Temperature Solar Hydrogen Production." *Chemical Reviews* 107(10): 4048–77.

Kodama, T., N. Gokon, and R. Yamamoto. 2008. "Thermochemical Two-Step Water Splitting by ZrO2-Supported NixFe3−xO4 for Solar Hydrogen Production." *Solar Energy* 82(1): 73–9. doi:http://dx.doi.org/10.1016/j.solener.2007.03.005.

Kodama, T., S. Enomoto, T. Hatamachi, and N. Gokon. 2008. "Application of an Internally Circulating Fluidized Bed for Windowed Solar Chemical Reactor with Direct Irradiation of Reacting Particles." *Journal of Solar Energy Engineering* 130(1): 014504-014504-4. doi:10.1115/1.2807213.

Kodama, T., T. Hasegawa, A. Nagasaki, and N. Gokon. 2009. "A Reactive Fe-YSZ Coated Foam Device for Solar Two-Step Water Splitting." *Journal of Solar Energy Engineering* 131(2): 021008. doi:10.1115/1.3090819.

Kodama, T., T. Shimizu, T. Satoh, M. Nakata, and K. I. Shimizu. 2002. "Stepwise Production of CO-Rich Syngas and Hydrogen via Solar Methane Reforming by Using a Ni(II)-Ferrite Redox System." *Solar Energy* 73(5): 363–74. doi:10.1016/S0038-092X(02)00112-3.

Kodama, T., Y. Kondoh, R. Yamamoto, H. Andou, and N. Satou. 2005. "Thermochemical Hydrogen Production by a Redox System of ZrO2-Supported Co(II)-Ferrite." *Solar Energy* 78(5): 623–31. doi:http://dx.doi.org/10.1016/j.solener.2004.04.008.

Kodama, T., Y. Nakamuro, and T. Mizuno. 2006. "A Two-Step Thermochemical Water Splitting by Iron-Oxide on Stabilized Zirconia." *Journal of Solar Energy Engineering* 128(1): 3–7.

Konstandopoulos, G., and C. Agrofiotis. 2006. "Hydrosol : Advanced Monolithic Reactors for Hydrogen Generation from Solar Water Splitting." *Revue Des Energies Renouvelables* 9: 121–26.

Kothari, R., D. Buddhi, and R. L. Sawhney. 2008. "Comparison of Environmental and Economic Aspects of Various Hydrogen Production Methods." *Renewable and Sustainable Energy Reviews* 12(2): 553–63.

Kubo, S., H. Nakajima, S. Kasahara, S. Higashi, T. Masaki, H. Abe, and K. Onuki. 2004. "A Demonstration Study on a Closed-Cycle Hydrogen Production by the Thermochemical Water-Splitting Iodine–sulfur Process." *Nuclear Engineering and Design* 233(1–3): 347–54. doi:http://dx.doi.org/10.1016/j.nucengdes.2004.08.025.

Laguna-Bercero, M. A. 2012. "Recent Advances in High Temperature Electrolysis Using Solid Oxide Fuel Cells: A Review." *Journal of Power Sources* 203: 4–16.

Malavasi, L., C. A. J. Fisher, and M. Saiful Islam. 2010. "Oxide-Ion and Proton Conducting Electrolyte Materials for Clean Energy Applications: Structural and Mechanistic Features." *Chemical Society Reviews* 39(11): 4370–87.

Meier, A., and A. Steinfeld. 2008. "Solar Energy in Thermochemical Processing." *Encyclopedia of Sustainability Science and Technology*, 9588–619. doi:10.1007/978-1-4614-5806-7_689.

Nakamura, T. 1977. "Hydrogen Production from Water Utilizing Solar Heat at High Temperatures." *Solar Energy* 19(5): 467–75. doi:http://dx.doi.org/10.1016/0038-092X(77)90102-5.

Perkins, C. Ã., P. R. Lichty, and A. W. Weimer. 2008. "Thermal ZnO Dissociation in a Rapid Aerosol Reactor as Part of a Solar Hydrogen Production Cycle." 33: 499–510. doi:10.1016/j.ijhydene.2007.10.021.

Perret, R. 2011. "Solar Thermochemical Hydrogen Production Research (STCH) Thermochemical Cycle Selection and Investment Priority." *Sandia Report*, May: 1–117.

Pohlenz, J. B., and N. H. Scott. 1966. "Method for Hydrogen Production by Catalytic Decomposition of a Gaseous Hydrocarbon Stream." Google Patents. http://www.google.com/patents/US3284161.

Pregger, T., D. Graf, W. Krewitt, C. Sattler, M. Roeb, and S. Möller. 2009. "Prospects of Solar Thermal Hydrogen Production Processes." *International Journal of Hydrogen Energy* 34(10): 4256–67. doi:http://dx.doi.org/10.1016/j.ijhydene.2009.03.025.

Ramachandran, R., and R. K. Menon. 1998. "An Overview of Industrial Uses of Hydrogen." *International Journal of Hydrogen Energy* 23(7): 593–98. doi:http://dx.doi.org/10.1016/S0360-3199(97)00112-2.

Roeb, M., C. Sattler, R. Klüser, N. Monnerie, L. de Oliveira, A. G. Konstandopoulos, C. Agrafiotis, et al. 2006. "Solar Hydrogen Production by a Two-Step Cycle Based on Mixed Iron Oxides." *Journal of Solar Energy Engineering* 128(2): 125. doi:10.1115/1.2183804.

Roeb, M., D. Thomey, L. de Oliveira, C. Sattler, G. Fleury, F. Pra, P. Tochon, et al. 2013. "Sulphur Based Thermochemical Cycles: Development and Assessment of Key Components of the Process." *International Journal of Hydrogen Energy* 38(14): 6197–204. doi:http://dx.doi.org/10.1016/j.ijhydene.2013.01.068.

Roeb, M., J.-P. Säck, P. Rietbrock, C. Prahl, H. Schreiber, M. Neises, L. de Oliveira, et al. 2011. "Test Operation of a 100 kW Pilot Plant for Solar Hydrogen Production from Water on a Solar Tower." *Solar Energy* 85(4): 634–44. doi:http://dx.doi.org/10.1016/j.solener.2010.04.014.

Roeb, M., M. Neises, N. Monnerie, F. Call, H. Simon, C. Sattler, M. Schmücker, and R. Pitz-Paal. 2012. "Materials-Related Aspects of Thermochemical Water and Carbon Dioxide Splitting: A Review." *Materials* 5(11): 2015–54. doi:10.3390/ma5112015.

Rubbia, C., and D. Salmieri. 2012. "Thermal Cracking of Methane into Hydrogen for a CO 2-Free Utilization of Natural Gas." 8: 4–9.

Scott, D. S. 2008. *Smelling Land: The Hydrogen Defense against Climate Catastrophe*. Victoria, B.C: Queen's Printer Publishing.

Shafiee, S., and E. Topal. 2009. "When Will Fossil Fuel Reserves Be Diminished?" *Energy Policy* 37(1): 181–89. doi:http://dx.doi.org/10.1016/j.enpol.2008.08.016.

Sheu, E. J., E. M. A. Mokheimer, and A. F. Ghoniem. 2015. "A Review of Solar Methane Reforming Systems." *International Journal of Hydrogen Energy* 40(38): 12929–55. doi:10.1016/j.ijhydene.2015.08.005.

Shindo, Y., N. Ito, K. Haraya, T. Hakuta, and H. Yoshitome. 1984. "Kinetics of the Catalytic Decomposition of Hydrogen Iodide in the Thermochemical Hydrogen Production." *International Journal of Hydrogen Energy* 9(8): 695–700. doi:http://dx.doi.org/10.1016/0360-3199(84)90267-2.

Siegel, N. P., J. E. Miller, I. Ermanoski, R. B. Diver, and E. B. Stechel. 2013. "Factors Affecting the Efficiency of Solar Driven Metal Oxide Thermochemical Cycles." *Industrial & Engineering Chemistry Research* 52(9): 3276–86. doi:10.1021/ie400193q.

Spiewak, I., C. E. Tyner, and U. Langnickel. 1993. "Applications of Solar Reforming Technology." 54. doi:10.2172/10131314.

Steinfeld, A. 2005. "Solar Thermochemical Production of Hydrogen—a Review." *Solar Energy* 78(5): 603–15. doi:http://dx.doi.org/10.1016/j.solener.2003.12.012.

Steinfeld, A., and A. W. Weimer. 2010. "Thermochemical Production of Fuels with Concentrated Solar Energy." *Optics Express* 18(9): A100–111. doi:10.1364/OE.18.00A100.

Steinfeld, A., C. Zurich, and R. Palumbo. 2001. "Solar Thermochemical Process Technology." *Encyclopedia of Physical Science & Technology* 15: 237–56. doi:10.1016/B0-12-227410-5/00698-0.

Stojić, D. L., M. P. Marčeta, S. P. Sovilj, and Š. S. Miljanić. 2003. "Hydrogen Generation from Water Electrolysis—possibilities of Energy Saving." *Journal of Power Sources* 118(1/2): 315–19. doi:10.1016/S0378-7753(03)00077-6.

Sub, H., Y. Ho, S. Jin, C. Sik, K. Kwang, and J. Gyu. 2012. "Continuous Bunsen Reaction and Simultaneous Separation Using a Counter-Current Flow Reactor for the Sulfur E Iodine Hydrogen Production Process." *International Journal of Hydrogen Energy* 38(14): 6190–6196.

Sugarmen, C., A. Rotstein, U. Fisher, and J. Sinai. 2004. "Modification of Gas Turbines and Operation with Solar Produced Syngas." *Journal of Solar Energy Engineering* 126(3): 867–71. http://dx.doi.org/10.1115/1.1758725.

Summers, W. A., and M. R. Buckner. 2008. "Hybrid Sulfur Thermochemical Process Development Objectives for FY 2005." 323–28. https://www.hydrogen.energy.gov/pdfs/progress05/iv_g_7_summers.pdf.

Takai, T., S. Kubo, T. Nakagiri, and Y. Inagaki. 2011. "Lab-Scale Water-Splitting Hydrogen Production Test of Modified Hybrid Sulfur Process Working at around 550°C." *International Journal of Hydrogen Energy* 36(8): 4689–701. doi:http://dx.doi.org/10.1016/j.ijhydene.2011.01.081.

Tamme, R., R. Buck, M. Epstein, U. Fisher, and C. Sugarmen. 2001. "Solar Upgrading of Fuels for Generation of Electricity." *Journal of Solar Energy Engineering* 123(2): 160–63. http://dx.doi.org/10.1115/1.1353177.

Tsai, C., K. Lee, J. S. Yoo, X. Liu, H. Aljama, L. D. Chen, C. F. Dickens, T. S. Geisler, C. J. Guido, and T. M. Joseph. 2016. "Direct Water Decomposition on Transition Metal Surfaces: Structural Dependence and Catalytic Screening." *Catalysis Letters* 146(4): 718–24.

Turner, J. A. 2004. "Sustainable Hydrogen Production." *Science* 305(5686): 972–74.

Udagawa, J., P. Aguiar, and N. P. Brandon. 2007. "Hydrogen Production through Steam Electrolysis: Model-Based Steady State Performance of a Cathode-Supported Intermediate Temperature Solid Oxide Electrolysis Cell." *Journal of Power Sources* 166(1): 127–36. doi:http://dx.doi.org/10.1016/j.jpowsour.2006.12.081.

Wong, B., R. T. Buckingham, L. C. Brown, B. E. Russ, G. E. Besenbruch, A. Kaiparambil, R. Santhanakrishnan, and A. Roy. 2007. "Construction Materials Development in Sulfur–iodine Thermochemical Water-Splitting Process for Hydrogen Production." *International Journal of Hydrogen Energy* 32(4): 497–504. doi:http://dx.doi.org/10.1016/j.ijhydene.2006.06.058.

Z'Graggen, A., P. Haueter, D. Trommer, M. Romero, J. C. de Jesus, and A. Steinfeld. 2006. "Hydrogen Production by Steam-Gasification of Petroleum Coke Using Concentrated Solar Power-II Reactor Design, Testing, and Modeling." *International Journal of Hydrogen Energy* 31(6): 797–811. doi:10.1016/j.ijhydene.2005.06.011.

Z'Graggen, A., P. Haueter, G. Maag, A. Vidal, M. Romero, and A. Steinfeld. 2007. "Hydrogen Production by Steam-Gasification of Petroleum Coke Using Concentrated Solar Power-III. Reactor Experimentation with Slurry Feeding." *International Journal of Hydrogen Energy* 32(8): 992–96. doi:10.1016/j.ijhydene.2006.10.001.

Zhenga, R., R. Diverb, D. Caldwella, B. Fritza, R. Camerona, P. Humblea, and R. Daglea. 2015. "Integrated Solar Thermochemical Reaction System for Steam Methane Reforming." *Energy Procedia* 69: 1192–200.

6

Solar Bio-Hydrogen Production: An Overview

Asheesh Kumar Yadav, Sanak Ray, Pratiksha Srivastava, and Naresh Kumar

CONTENTS

6.1 Introduction

The ever-increasing population and rapid growth of technology are leading to the increase in the consumption of fossil fuel (nonrenewable) at a higher rate compared with that of its production. Most of the energy today is derived from fossil fuels such as coal, oil shale, tar sands, petroleum, bitumen, and natural gas. Consequently, burning of fossil fuels is leading to major environmental problems like climate change, ozone layer depletion, global warming, to name a few (Jacobson 2009). Various treaties and conventions have been undertaken over the decades to protect environment globally and have recently reached at an agreement to reduce the consumption of fossil fuels. Further, the depleting fossil fuels tantalized various nations to focus on sustainable renewable alternatives. So, it is expected that renewable energy sources like biomass, hydropower, wind energy, solar energy, geothermal energy, tidal energy, and hydrogen will play significant role in fulfilling the energy needs of the future globally. European Union's energy policy predicts that by the year 2050, 50% of world energy requirements will be fulfilled by renewable resources, and nearly 80% of electricity will be produced from renewable source by 2040 (Wilkes et al. 2011). Therefore, it warrants finding more alternative energy sources and developing technologies that can replace the fossil fuels as an energy source in the coming decades.

Hydrogen is expected to play a key role as an energy vector. Presently, hydrogen and fuel cells are believed to be the most promising renewable sources of energy. Hydrogen contains highest energy per unit mass and produces environment friendly H_2O upon combustion. Presently, 400–500 billion Nm^3 of hydrogen is consumed globally per annum (Demirbas 2009). Currently, a large fraction of hydrogen produced is used by different industrial sectors like food, electronics, petrochemicals, and metallurgical, whereas a smaller fraction is used in the energy segment. Literature shows that hydrogen has been produced for a long time using conventional methods utilizing fossil fuels, leading to profound greenhouse gas emissions. Currently, it is being produced from hydrogen-rich feedstock such as water, biomass, or fossil fuel. Globally, 96% of hydrogen is produced using fossils (Abánades 2012), including 48% from steam methane reforming (SMR), 30% from oil or naphtha reforming from refineries or chemical industries, 18% from coal gasification, and the rest 3.9% from water electrolysis (Baghchehsaree et al. 2010).

Various conventional methods to produce hydrogen from feedstock also include thermochemical, electrolytic, and photolytic processes. Thermochemical processes (SMR, partial oxidation, and gasification) use heat and pressure for breaking hydrogen bonds. In electrolytic processes (water electrolysis), electricity is used to decompose water, forming O_2 and H_2 as end products, whereas in photolytic processes, H_2 is extracted from waste gases generated from algae (Padro and Putsche 1999).

In the view of above discussion, it is important to find out the lacunas in conventional processes and adopt biological processes that especially use vastly available solar radiation (solar energy) to overcome these gaps and produce biologically derived hydrogen, commonly known as bio-hydrogen. Nature has evolved microorganisms in such a way that they produce H_2 during the course of metabolism using the solar energy (light). An overview of such solar bio-hydrogen generation processes, on fundamentals, and, most importantly, the technological advancements made in this area so far are presented in this chapter.

6.2 Fundamentals of Bio-Hydrogen Production

6.2.1 Fundamentals

Metabolism is the fundamental of all life, which is a complex web of the flow of electrons. The need of metabolism is fulfilled from the breaking of chemical bonds or, in the case of photosynthetic organisms by the *light*. The light energy can be used by plants, algae, cyanobacteria, and photosynthetic bacteria to excite the electrons into higher energy states. These excited electrons are transported via electron transport chain and subsequently their reducing power is used by the organism to drive a variety of chemical reactions. Water is used as the source of these electrons in the case of plants, algae, and cyanobacteria. During the extraction of excited electrons from the water, water splits into oxygen and protons. Therefore, this form of photosynthesis is termed as oxygenic photosynthesis. Various bacteria like green sulfur (GS) and purple nonsulfur (PNS) bacteria and so on can also stimulate anoxygenic photosynthesis. In anoxygenic photosynthesis, electrons come from organic substrates; here, light is used to excite these electrons to higher energy, allowing them to carry out additional energy-yielding reactions, as these molecules are ultimately oxidized to CO_2. During photosynthesis, protons are generated and used to sustain a pH gradient, where movement of protons is coupled with the production of adenosine triphosphate (ATP). Hydrogen production in microorganisms is carried out by enzymes capable of reducing free protons to molecular hydrogen (H_2). Such enzymes include the uptake and reversible hydrogenases and the nitrogenases. There are also other biochemical processes attached with these enzymes along with hydrogen production. The energy is used by these enzymes in multiple steps from an organism's central energy inputs (photosynthesis or oxidative phosphorylation) and is given in the form of electron carriers such as ferredoxin or nicotinamide adenine dinucleotide phosphate (NADPH) and energy-yielding molecules like ATP.

There is a remarkable diversity of sequences in these enzymes, as there are several kinds of co-factors they use, and there are also many differences in how they are regulated by and integrated into the metabolism of the organisms in which they are found. The photosynthesis mechanism is the starting point of all biological solar-driven H_2 production (Rupprecht et al. 2006). The solar bio-hydrogen is generated in several biological processes that include oxygenic photosynthesis, anoxygenic photosynthesis, fermentation, and cyanobacterial metabolism. Figure 6.1 depicts the said processes. The oxygenic photosynthesis is light dependent and occurs in cyanobacteria and plants. The solar energy is the source of energy for plant and caynobacteria. Antenna proteins associated with photosystem-I (PSI) and photosystem-II (PSII) capture the solar energy. In case of favorable conditions to linear

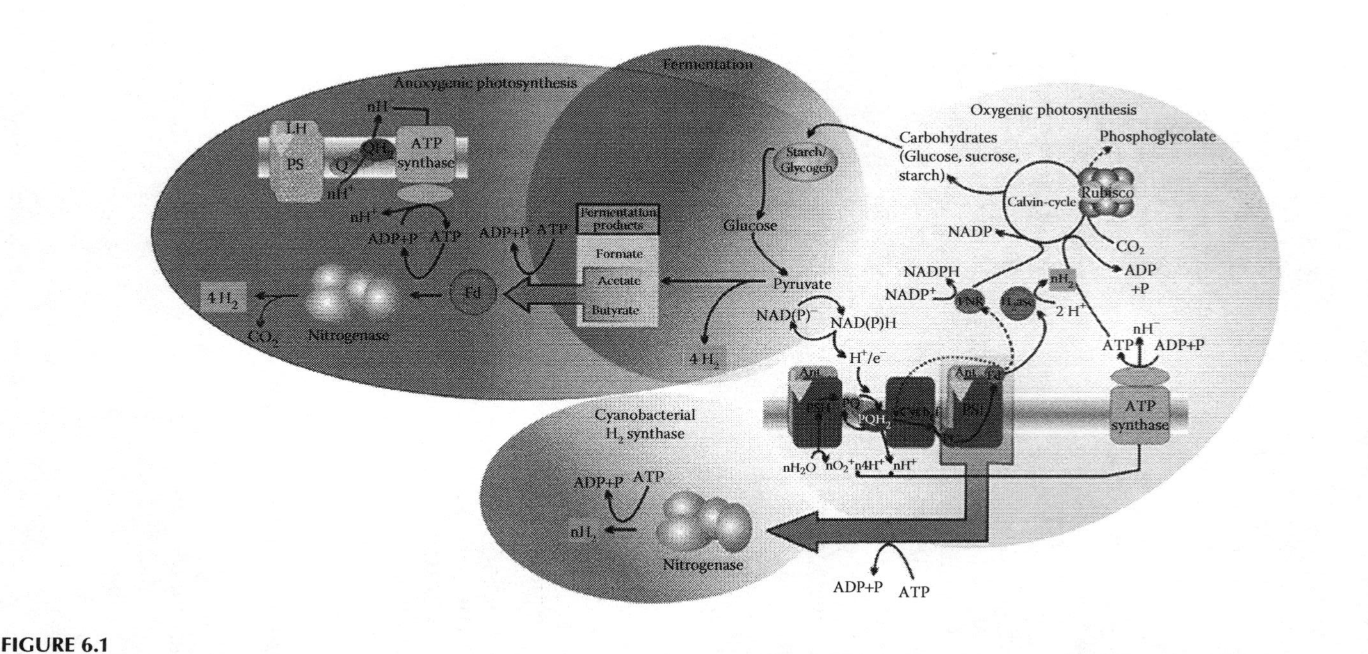

FIGURE 6.1
Photobiohydrogen production pathways: oxygenic photosynthesis, anoxygenic photosynthesis, fermentation, and cyanobacterial metabolism. (With kind permission from Springer Science + Business Media: *Applied Microbiology Biotechnology*, Perspectives and advances of biological H2 production in *microorganisms*, 72, 2006, 442–449, Rupprecht, J. et al.)

electron transport, electron extracted from H_2O is transported to PSII. From PSII, electrons travel along the photosynthetic electron transport chain via plastoquinone (PQ), the cytochrome b6f complex (Cyt b6f), plastocyanin (PC), photosystem I (PSI), and ferredoxin (Fd) before ending up in the production of NADPH by ferredoxin-NADP+ oxidoreductase.

Concurrently, PSII releases the H^+ into the thylakoid lumen and which takes part in PQ/PQH_2 cycle to create a gradient. At this point, ATP synthase assists the ATP production. In around PSI, electrons from the electron transport chain travel in a cyclic manner and used by synthesized ATP and NADPH for CO_2 fixation (Calvin cycle). Rubisco activity is accountable for integration of CO_2 into ribulosebis-phosphate, which is subsequently used to generate C3 and C6 sugars and, ultimately, starch. In the first step of photorespiration, rubisco catalyzes the oxygenation reaction. Certain algae under anaerobic conditions can use starch as a source of electron and proton for H_2 production using hydrogenase HydA enzyme.

In cyanobacteria, the H^+ and e^- extracted from H_2O are converted to H_2 by a nitrogenase or fermentation. Both pathways are ATP dependent. Some of the bacteria follow fermentation processes for producing H_2. Such bacteria convert carbohydrate into sugars and consequently pyruvate via glycolysis, which ultimately is converted into H_2 and organic acids. Anoxygenic photosynthesis is a phototrophic process, where light energy is captured and converted to ATP. In anoxygenic photosynthesis, oxygen is not produced and water is not used as electron donor. Photosynthetic PNS and GS bacteria can do anoxygenic photosynthesis.

6.2.2 Bio-Hydrogen Production Processes

Broadly, there are two major groups of bio-hydrogen production processes: light-dependent (or photobiological processes) and light-independent processes (Das and Veziroglu 2001; Levin et al. 2004; Show et al. 2012). Figure 6.2 depicts various bio-hydrogen production processes.

6.2.2.1 Light-Dependent (Photobiological) Bio-Hydrogen Production

Depending on the nature of reactions and microbes, light-dependent process can be grouped into direct and indirect biophotolysis and photofermentation.

6.2.2.1.1 Direct and Indirect Bio-Photolysis

Microalgae are well known for their photobiological hydrogen production and have received much scientific interest in the last few decades (Prince and Kheshgi 2005). In the direct biophotolysis, PSII and PSI absorb light energy simultaneously, which facilitates the transfer of electrons linearly from water to ferredoxin, and, consequently, reduced ferredoxin molecule acts as an electron donor for hydrogenase enzyme, which reversibly catalyzes the reduction of protons to hydrogen (Melis et al. 2000).

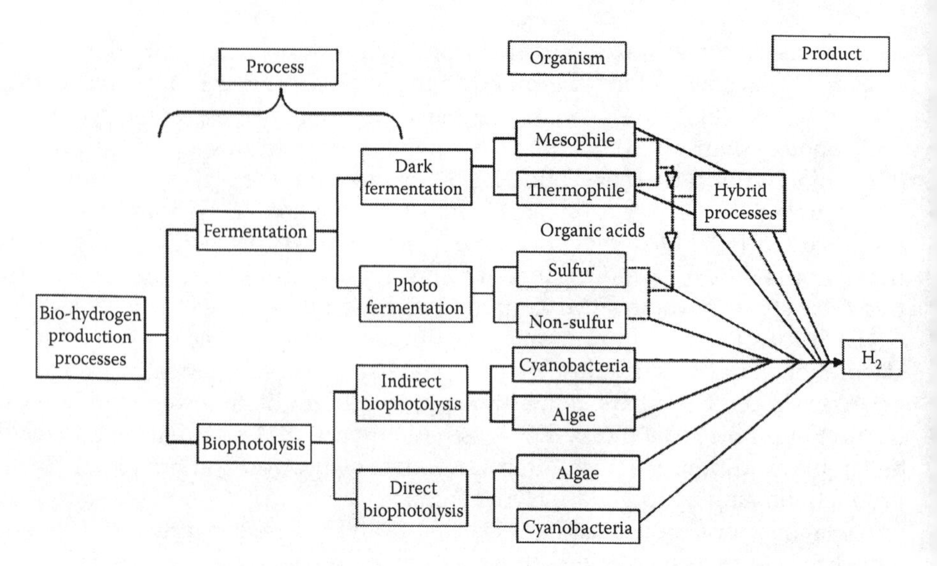

FIGURE 6.2
Bio-hydrogen production processes. (Khanna, N., Das, D.: Biohydrogen production by dark fermentation. *WIREs Energy Environment*. 2013, 2, 401–421. Copyright Wiley-VCH Verlag GmbH & Co. KGaA. Reproduced with permission.)

$$2H^+ + 2FD^- \leftrightarrow H_2 + 2FD \tag{6.1}$$

Green microalgae are well known for their very active [FeFe]-hydrogenase enzyme. It has relatively high 12%–14% solar-to-H_2 energy conversion efficiency (Melis 2009) and the ability to both oxidize water and generate H_2 in their chloroplast (Melis 2002; Melis and Happe 2001). The process is promising, as it produces hydrogen using water as substrate and solar energy. Studies have shown that oxygen produced during the photosynthetic process inhibits hydrogenase activity, resulting in decreased hydrogen production (Levin et al. 2004). In addition, the downside of direct biophotolysis is that it generates highly explosive H_2–O_2 mixtures.

The indirect biophotolysis occurs in two steps: in the first stage, the mixture of green algae and cyanobacteria is used, where green algae assimilate CO_2 into carbohydrate and cyanobacteria generate glycogen. In the second stage, stored carbohydrates produce H_2 catalyzed by reversible hydrogenases (Benemann 1997; Gupta et al. 2013; Levin et al. 2004; Zabrosky 1997).

These two phases are separated from each other, thereby ruling out the possibility of the inhibition of hydrogenase activity by oxygen and also avoiding explosive mixture formation (Prince and Kheshgi 2005). The two-stage indirect biophotolysis process can be explained by the following reactions:

$$6H_2O + 6CO_2 \xrightarrow{\text{Light energy}} C_6H_{12}O_6 + 6O_2 \tag{6.2}$$

$$C_6H_{12}O_6 + 6H_2O \xrightarrow{\text{Light energy}} 6CO_2 + 12H_2 \quad (6.3)$$

Cyanobacteria that produce H_2 are predominantly filamentous and nitrogen fixing and belong to the genus *Nostoc, Anabaena, Calothrix, Oscillatoria*. On the other hand, cyanobacteria that produce hydrogen are non-nitrogen fixing and belong to the genus *Synechocystis, Synechococcus, Gloeobacter* (Das and Veziroglu 2008).

6.2.2.1.2 *Photofermentation*

In the photofermentation, H_2 is produced from organic substrate (acids) by a diverse group of anaerobic photosynthetic bacteria in the presence of light. The advantages of this process include (1) high substrate conversion rate, (2) operating anaerobically (ruling out the oxygen sensitivity issue that adversely affects the hydrogenase, nitrogenase enzymes), (3) use of wider wavelength of light, and (4) ability to use waste organic (Das and Veziroglu 2001; Gupta et al. 2013).

The PNS, a photosynthetic microorganism, appear more promising, as they require less free energy to decompose organic substrates (Basak and Das 2007). PNS generate H_2 as a byproduct of nitrogenase activity, induced under nitrogen-deficiency conditions, that is facilitated by sunlight as the energy source and small organic molecules as the carbon substrate (Eroglu and Melis 2011).

6.2.2.2 Light-Independent (Dark Fermentation) Bio-Hydrogen Production

Dark fermentative bacteria are the heterotrophs that produce hydrogen in anoxic conditions using organic sources for their growth. In the dark fermentation, hydrogen production takes place in two catabolic steps: first is the decarboxylation of pyruvate into acetyl-CoA (Equation 6.4), which generates reduced ferredoxin (Equation 6.5), a direct electron donor for hydrogenase (Khanna and Das 2013):

$$\text{Pyruvate} + \text{CoA} + 2\text{Fd}(\text{ox}) \rightarrow \text{Acetyl CoA} + 2\text{Fd}(\text{red}) + CO_2 \quad (6.4)$$

$$2\text{Fd}(\text{red}) + 2H^+ \rightarrow 2\ \text{Fd}(\text{ox}) + H_2 \quad (6.5)$$

In the second stage, pyruvate and Co-A are converted into formate and acetyl-CoA. Subsequently, formate produces CO_2 and H_2 catalyzed by the enzyme formate hydrogen lyase (FHL).

$$\text{Pyruvate} + \text{CoA} \rightarrow \text{Formate} + \text{Acetyl CoA} \quad (6.6)$$

$$\text{Formate} \rightarrow CO_2 + H_2 \quad (6.7)$$

$$2H^+ + 2e^- \rightarrow H_2 \quad (6.8)$$

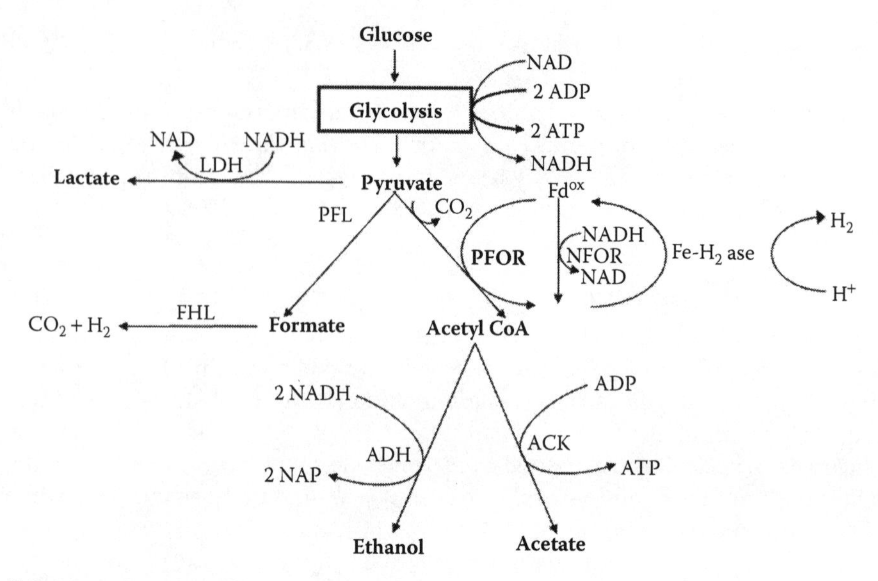

FIGURE 6.3
Overall biochemical pathway for dark fermentative metabolism (LDH, lactate dehydrogenase; PFOR, pyruvate ferredoxin oxydoreductase; PFL, pyruvate ferredoxin oxydoreductase; PFL, pyruvate formatelyase; FHL, formate hydrogen lyase; ADH, alcohol dehydrogenase; ACK, acetyl CoA kinase; NFOR-NADH, ferredoxin oxydoreductase.) (Khanna, N., and D. Das, Biohydrogen production by dark fermentation. *WIREs Energy Environment.* 2013, 2, 401–421. Copyright Wiley-VCH Verlag GmbH & Co. KGaA. Reproduced with permission.)

This is the general mechanism of H_2 generation in facultative anaerobes like *Klebsiella* sp. (Khanna and Das 2013). The overall biochemical pathway for dark fermentation is illustrated in Figure 6.3.

6.3 State of the Art

6.3.1 Photobiohybrid for Hydrogen Production

In recent years, an interesting approach known as "photobiohybrid" has emerged. Photobiohybrid can be defined as a device that integrates biocatalyst with light-harvesting nanomaterials. Various kinds of photobiohybrids designed to produce H_2 are shown in Figure 6.4. One of the early examples of light-driven H_2 production by a photobiohybrid complex was reported in the 1980s using [FeFe]-hydrogenase coupled with nanoparticulate TiO_2 (Cuendet et al. 1986). Later efforts integrated [NiFe]-hydrogenases with anatase TiO_2 (Pedroni et al. 1996; Selvaggi et al. 1999) and CdS (Shumilin et al. 1992). More recently, nanoparticulate TiO_2 were sensitized with dye molecules and [NiFeSe]-hydrogenases to drive H_2 production (Reisner et al. 2009a,b).

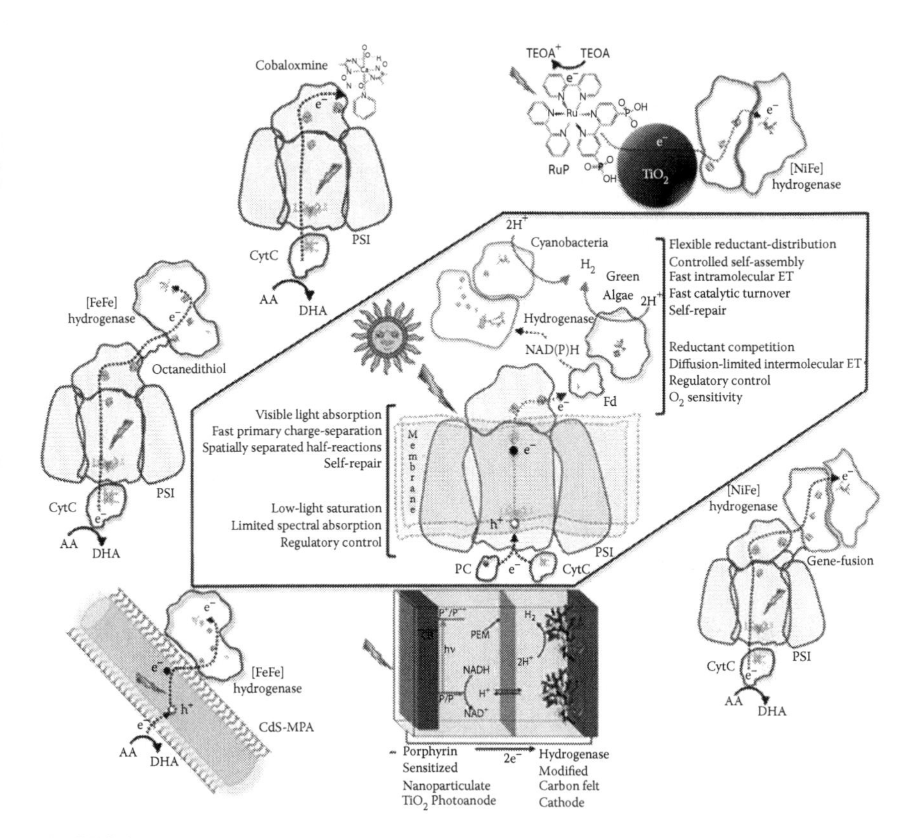

FIGURE 6.4
Photobiohybrid complexes and devices directly couple the catalytic power of enzymes and catalysts with light harvesting by natural photosystems or artificial nanoparticles and photoelectrochemical cells for solar H_2 production. (With kind permission from Springer Science + Business Media: *Microbial BioEnergy: Hydrogen Production, Advances in Photosynthesis and Respiration*, 38, 2014, 101–135, Ghirardi et al.) Clockwise from the left: a PSI-CytC fusion chemically linked to a [FeFe]-hydrogenase by octanedithiol (labeled) (Data from Lubner, C.E. et al., *Proceeding of the National Academy of Sciences* USA, 108, 20988–20991, 2011); cobaloxime adsorbed onto PSI (Utschig, L.M. et al., *Journal of the American Chemical Society*, 133, 16334–16337, 2011.); [NiFeSe]-hydrogenase adsorbed to particulate, dye-sensitized TiO_2 (Data from Reisner, E. et al., *Chemical Communications*, 2009, 550–552, 2009b.); [NiFe]-hydrogenase genetically fused to the *psaE* subunit to create a PSI-[NiFe]-hydrogenase complex (Data from Ihara, M. et al., *Photochemistry and Photobiology*, 82, 1677–1685, 2006a; Ihara et al. 2006b; Krassen, H. et al., *ACS Nano*, 3, 4055–4061, 2009.); [FeFe]-hydrogenase integrated into a dye-sensitized photoelectrochemical cell (Reprinted with permission from Hambourger, M. et al., *Journal of the American Chemical Society*, 130, 2015–2022, 2008.); and [FeFe]-hydrogenase adsorbed to mercapto-propionic acid capped CdS nanorods (Data from Brown et al. 2012). AA, ascorbic acid; CytC, cytochrome C; CdS, cadmium sulfide nanorod; DHA, dehydroascorbate; ET, electron transfer; MPA, mercaptopropionic acid; PC, plastocyanin; RuP, ruthenium bipyridinephosphonic acid; TEOA, triethanolamine.

With time, an advancement in material synthesis research for improved, tunable control of band-gap energy, band-edge positions, and surface functionalization (Gaponik et al. 2002) and an enhancement in the understanding of the properties like control self-assembled layers on both nanoparticles and electrodes (Badia et al. 2000; Zhang et al. 2011) have provided opportunity to design better artificial molecular system (Allakhverdiev et al. 2009; Gust et al. 2000, 2009; Lubitz et al. 2008; Navarro et al. 2010; Park and Holt 2010; Reisner 2011; Wang et al. 2011). Still, there is need for further progress in photobiohybrids research as measured turnover numbers and rates are far below than those obtained *in vivo*. Nevertheless, these attempts are very promising toward developing systems for investigating the mechanisms of the PSII-oxygen-evolving complex water splitting reaction. Various enzymes including hydrogenases are ideal biocatalysts and can be operated near the thermodynamic potential of the specific half-reaction with high turnover (Armstrong et al. 2009; Armstrong and Hirst 2011; Cracknell et al. 2008). Madden et al. (2012) reported that algal and bacterial [FeFe]-hydrogenases can catalyze H_2 evolution at k_{cat} values of up to 10^4 s^{-1}, with long-standing stabilities when immobilized on electrode surfaces (Alonso-Lomillo et al. 2007; Bae et al. 2008; Hambourger et al. 2008). In addition, [NiFeSe]-hydrogenases can evolve H_2 at high turnover ($k_{cat} \sim 10^3$ s^{-1}) and retain activity in the presence of O_2 (Baltazar et al. 2011; Reisner et al. 2009a; Vincent et al. 2006). Since hydrogenase turnover is fast, it is feasible to match them with the high light-harvesting capacity (k_{Abs} >> solar flux) and broad spectral range of artificial photochemical materials. In principle, charge-transfer efficiencies and H_2 production rates can approach or even surpass those of photosynthetic organisms (Blankenship et al. 2011; Bolton et al. 1985) but at the cost of a self-repair process. Therefore, considerable challenge before photobiohybrid devices to produce H_2 is to develop designs/integration that can attain a cautious balance of light absorption, conversion, stability, and catalysis efficiency.

6.3.2 Integrated Technologies for Hydrogen Production

The individual process cannot meet the desired hydrogen yield to the level of commercial viability, as of today. Such restrictions are due to thermodynamic limitations, for example, 1 mol of glucose can generate 12 mol of H_2, but complete oxidation of glucose into H_2 is not possible, as the reaction is not thermodynamically favorable.

$$C_6H_{12}O_6 + 6H_2O \rightarrow 12H_2 + 6CO_2, \quad G_0 = +3.2\ \text{kJ} \tag{6.9}$$

To overcome the thermodynamic barriers, merging of two processes may lead to promising outcomes. There are some attempts to integrate two biological processes, where first process produces the substrate for the organism that is used for the second process to enhance the efficiency of harvesting solar energy (light) or utilization of organic compounds (Tekucheva and Tsygankov 2012). This section describes some of the practical and hypothetical integrations for maximizing the hydrogen production efficiency.

6.3.2.1 Dark Fermentation Process and Photofermentation

The end products of dark fermentation are reduced metabolites like acetic acid, butyric acid, and so on. These end products are not further useful for the microbes of the dark fermentation. However, photosynthetic bacteria are capable of utilizing such products by fixing nitrogen and producing hydrogen under anaerobic conditions. These bacteria can overcome the thermodynamic barrier in doing this. The major benefit of such a system is that this may increase the theoretically possible H_2 yield to 12 mol/mol glucose (Das and Veziroglu 2001). Currently, this method is generally used for generating H_2. The overall reactions in an integrated dark and photofermentation system are given below (Das 2009).

Dark fermentation stage:

$$C_6H_{12}O_6 + 2H_2O \rightarrow 2CH_3COOH + 2CO_2 + 4H_2 \tag{6.10}$$

Photofermentation stage:

$$2CH_3COOH + 4H_2O \rightarrow 4CO_2 + 8H_2 \tag{6.11}$$

The following factors should be taken into account when selecting a substrate for such a two-stage system (Tekucheva and Tsygankov 2012):

1. The organic substrates contained in the waste should be fermented during the dark stage to produce by-products that can be readily utilized by purple bacteria.
2. The nitrogenase activity of purple bacteria should not be inhibited or decreased by an excess concentration of nitrogenous compounds contained in the dark bioreactor effluent after the first stage.
3. The presence and concentration of mineral components in the waste should not limit or inhibit the hybrid processes.

While selecting the bacteria for the two-stage process, it is essential to take into account the following four combinations of bacterial metabolic pathways, which should enable to achieve the theoretical yield of 12 mol H_2/mol glucose/hexose (Redwood et al. 2009):

1. 1 glucose $\rightarrow$ 2 lactates + 0 H_2 (lactic fermentation) and
   2 lactates $\rightarrow$ 12 H_2 (purple bacteria);

2. 1 glucose → 4 H_2 + 2 acetates (acetic fermentation) and 2 acetates → 8 H_2(purple bacteria);
3. 1 glucose → 2 H_2 + 1 ethanol + 1 acetate (mixed fermentation) and 1 ethanol + 1 acetate → 10 H_2(purple bacteria); and
4. 1 glucose → 2 – 4 H_2 + 0 – 1 butyrate + 0 – 2 acetate(s) (acetobutyric fermentation) and 1 butyrate → 8 – 10 H_2(purple bacteria).

6.3.2.2 Dark Fermentation, Biophotolysis, and Photofermentation

This combination of bioprocesses is more of a theoretical approach and has received limited attention so far. In this approach, microalgae can also be used to generate carbohydrates as substrate for H_2 generation. This substrate can further be utilized for dark fermentation. The generated end metabolites and CO_2 in dark fermentation are further utilized in photofermentation and biophotolysis process to produce hydrogen. This type of integration promises high hydrogen yields close to the theoretical maximum.

6.3.2.3 Biophotolysis and Photofermentation

Biophotolysis and photofermentation integration process works similar to the integration of dark fermentation with photofermentation. It involves two-stage process, where biophotolytic production of hydrogen by algae occurs in the first stage. Further, the used-up media well-off in organic acids are subjected to photofermentation. The bacteria of photofermentation utilize the substrate coming from first stage for generating hydrogen. The integrated process has been already demonstrated at the pilot scale by Miura et al. in 1997. They used tubular reactors for indirect biophotolysis using *Chlamydomonas* MGA 161. The buildup starch was fermented *in situ* to produce organic acids in the natural day and night cycle. The conversion yield of organic compounds from starch of *Chlamydomonas* MGA 161 was 80%–100% that of the theoretical yield. In a second step, photofermentation was carried out by *Rhodovulum sulfidophilum* W-1S from organic compounds produced by *Chlamydomonas* MGA 161. The molar yield of hydrogen photoproduction of *R. sulfidophilum* W-1S was about 40%.

6.3.2.4 Dark Fermentation and Anaerobic Treatment

Cavinato et al. (2012) have tested this type of integrated process at pilot level with the aim of integrating hydrogen production with methane production. In this approach, bio-hydrogen can be produced through dark fermentation in the first stage and the spent substrate can be used for biomethanation. This process is feasible because methane is produced by acetoclastic bacteria. The acetoclastic bacteria need organic acids, which are generally present in spent media after bio-hydrogen production. By this integrated process, mixture of hydrogen and natural gas is generated. This mixture is known as hythane. It contains about 20% hydrogen and 80% natural gas.

6.3.2.5 Microbial Electrolysis Cell

The microbial electrolysis cell (MEC) is an emerging technology that merges bacterial metabolism with electrochemistry to produce hydrogen. MEC technology provides the necessary energy to convert organic acids (present in the spent media) to hydrogen (Cheng and Logan 2007; Ditzig et al. 2007; Rozendal et al. 2008). The biggest advantage with MEC technology is that the microbes present in the anodic chamber can completely ferment glucose and near stoichiometric yields of hydrogen can be achieved.

MECs primarily utilize mixed consortia of microbes, which carry out *in-situ* dark fermentation, and the consequential fermentation products are used by the electrogenic members of the community (Cheng and Logan, 2007). In MECs, the electrogenic bacteria catabolize their substrate and the electrode (anode) acts as the terminal electron acceptor. Interestingly, to drive hydrogen production at the cathode, a supplementary voltage (>200 mV) is also provided in addition to the voltage generated from the oxidation of the substrate (Cheng and Logan 2007). Thus, in principle, a second-stage MEC after an initial fermentative hydrogen stage could completely convert a substrate to hydrogen, achieving 12 mol H_2/mol of glucose with only a bit of extra cost of electricity (Hallenbeck and Ghosh 2009). MEC is still under research phase and there has been continuous improvement. MEC is one of the most promising technologies and has the potential to produce H_2 at commercially viable scale.

6.4 Challenges and Strategies

There are three main immediate issues to be addressed to improve bio-hydrogen production: (1) low photochemical efficiencies, (2) sensitivity of hydrogenases to O_2, and (3) competition for reductant from ferredoxin between hydrogenases and other cellular functions (Brentner et al. 2010). Different photobioreactor designs are being used to address the first two issues, and genetic engineering is also being exploited to address all the three issues (Brentner et al. 2010). A photochemical efficiency of 10% is generally being targeted (Melis 2008; Mussgnug et al. 2007; Turner et al. 2008), but present efficiencies are far from touching this target. It is mainly due to limitations in light penetration within photobioreactors and the transfer of light energy within cells. High sensitivity of [FeFe]-hydrogenases to O_2 makes their isolation complicated and restricts the accessibility to characterize the enzyme's structure (Brentner et al. 2010). Many engineering efforts are

TABLE 6.1

Various Biological Hydrogen Production Processes with General Overall Reactions Involved Therein, Broad Classification of Microorganisms Used, and Their Challenges and Deployments

Process	General Reactions and Broad Classification of Microorganism Used	Challenges	Deployments
Dark fermentation	$C_6H_{12}O_6 + 2H_2O \rightarrow 2CH_3COOH + 2CO_2 + 4H_2$ Fermentative bacteria	• Relatively lower achievable yields of hydrogen • As yields increase, hydrogen fermentation becomes thermodynamically unfavorable • Product gas mixture contains CO_2, which has to be separated	• It can produce H_2 all day long without light • A variety of carbon sources can be used as substrates • It produces valuable metabolites such as butyric, lactic, and acetic acids as byproducts • It is anaerobic process, so there is no O_2 limitation problem
Photofermentation	$H_3COOH + 2H_2O + light \rightarrow 4H_2 + 2CO_2$ Purple bacteria, Microalgae	• Light conversion efficiency is very low, only 1%–5% • In-homogeneity of light distribution	A wide spectral light energy can be used by these bacteria • Can use different waste materials
Direct biophotolysis	$2H_2O + light \rightarrow 2H_2 + O_2$ Microalgae	• O_2 sensitivity of hydrogenase enzyme • Low light conversion efficiency	• Can produce H_2 directly from water and sunlight • Solar conversion energy increased by 10-fold as compared to trees, crops
Indirect biophotolysis	$6H_2O + 6CO_2 + light \rightarrow C_6H_{12}O_6 + 6O_2$ • $C_6H_{12}O_6 + 2H_2O \rightarrow 4H_2 + 2CH_3COOH + 2CO_2$ • $2CH_3COOH + 4H_2O + light \rightarrow 8H_2 + 4CO_2$ • Overall reaction: • $12H_2O + light \rightarrow 12H_2 + 6O_2$ • Microalgae, cyanobacteria	• Enzyme inhibition by O_2 • H_2 consumption by an uptake hydrogenase • Overall low production rates and yield	• Can produce H_2 from water • Has the ability to fix N_2 from atmosphere

Source: Khanna, N., Das, D.: Biohydrogen production by dark fermentation. *WIREs Energy Environment*. 2013, 2, 401–421. Copyright Wiley-VCH Verlag GmbH & Co. KGaA. Reproduced with permission.

therefore focused on the more accessible, less-O_2-sensitive, cyanobacterial [NiFe]-hydrogenases. A membrane-bound, soluble hydrogenase has been characterized in the bacterium *Ralstoniaeutropha*, which is relatively O_2 tolerant but less active than the O_2-sensitive hydrogenases (Rousset and Cournac 2008). Researchers are also attempting to improve the catalytic efficiency of [NiFe]-hydrogenases. One approach is the modification of intramolecular electron transfer pathways in bidirectional [NiFe]-hydrogenase to alter the redox potential at the active site for preferential catalysis in one direction (Rousset and Cournac 2008). Table 6.1 presents the summary of various bio-hydrogen processes, challenges, and deployments status.

6.5 Conclusion

The world energy demand is increasing, fossil fuel resources are depleting, and the need to minimize greenhouse gases is becoming a global priority. In such a scenario, bio-hydrogen production is one of the finest options to provide a sustainable solution to address the above-mentioned issues. Hydrogen will be one of the pragmatic energy options in future. It is a known fact that hydrogen is a clean and efficient energy carrier and can be produced by managing the renewable sources. Chemical methods of hydrogen production require high costs. Thus, low-cost hydrogen production alternative is urgently needed. In such a scenario, biological method has the potential for a substitute to the current renewable technologies. There are various biodegradable waste substrates that can be utilized for producing H_2 through biological route. There are various technologies used for biological hydrogen production like biophotolysis of water by cyanobacteria and green algae, photobiohybrid material-based technologies, photofermentation, dark fermentation, photodark fermentation, bioelectrochemical processes, and combination of these technologies. Research on photobiological hydrogen metabolism has increased significantly; further studies need to be more innovative to increase the effectiveness of photobioreactors. Hydrogen production using biophotolysis systems by cyanobacteria and green algae have the potential to contribute significantly in the renewable energy generation to meet the global energy demand. Biological processes like direct biophotolysis can produce H_2 directly from water. Although H_2 yield is relatively low, this process has provided new knowledge about the enzymes like hydrogenases, biomaterial, and the nature of electron carriers in the photosynthesis system.

Alternatively, indirect biophotolysis has its advantages and potential, where co-generation of H_2 involves steps of photosynthesis and dark fermentation of biomass. However, in the photofermentation, H_2 from organic compounds can be generated in the presence of light. Integration of two processes is also receiving interest to develop high hydrogen yielding process. MEC combines the electrochemistry and bio-hydrogen-generating process and leads to higher yield of H_2.

H_2 is the future energy, but, at present, it does not have sufficient share in global energy scenario. More research on how to improve the H_2 yields along with production rates for realistic economically feasible applications is required.

References

Abánades, A. 2012. The challenge of hydrogen production for the transition to CO_2-free economy. *Agronomy Research*, International Scientific Conference *Biosystems Engineering Special Issue 1*: 11–16. http://agronomy.emu.ee/vol10Spec1/p10s102.pdf.

Allakhverdiev, S.I., V.D. Kreslavski, V. Thavasi, S.K. Zharmukhamedov, V.V. Klimov, T. Nagata, H. Nishihara, S. Ramakrishna 2009. Hydrogen photoproduction by use of photosynthetic organisms and biomimetic systems. *Photochem Photobiol Sci 8*, no. 2 (February): 148–156. doi:10.1039/b814932a.

Alonso-Lomillo, M.A., Rudiger, O., Maroto-Valiente, A., Velez, M., Rodriguez-Ramos, I., Munoz, F.J., Fernandez, V.M., De Lacey, A.L. 2007. Hydrogenase-coated carbon nanotubes for efficient H_2 oxidation. *Nano Letters 7*, no. 6 (May): 1603–1608. doi:10.1021/nl070519u.

Armstrong, F.A., Belsey, N.A., Cracknell, J.A., Goldet, G., Parkin, A., Reisner, E., Vincent, K.A., Wait, A.F. 2009. Dynamic electrochemical investigations of hydrogen oxidation and production by enzymes and implications for future technology. *Chemical Society Reviews 38* (December): 36–51. doi:10.1039/B801144N.

Armstrong, F.A., Hirst, J. 2011.Reversibility and efficiency in electrocatalytic energy conversion and lessons from enzymes. *Proceedings of the National Academy of Sciences USA 108*, no. 34 (August): 14049–14054. doi:10.1073/pnas.1103697108.

Badia, A., Lennox, R.B., Reven, L. 2000. A dynamic view of self-assembled monolayers. *Accounts of Chemical Research 33*, no. 7 (March): 475–481. doi:10.1021/ar9702841.

Bae, S., Shim, E., Yoon, J., Joo, H. 2008. Photoanodic and cathodic role of anodized tubular titania in lightsensitized enzymatic hydrogen production. *Journal of Power Sources 185*, no. 1 (October): 439–444. doi:10.1016/j.jpowsour.2008.06.094.

Baghchehsaree, B., Nakhla, G., Karamanev, D., and Argyrios, M. 2010. Fermentative hydrogen production by diverse microflora. *International Journal of Hydrogen Energy 35*, no.10 (May): 5021–5027. doi:10.1016/j.ijhydene.2009.08.072.

Baltazar, C.S.A., Marques, M.C., Soares, C.M., DeLacey, A.M., Pereira, I.A.C., Matias, P.M. 2011. Nickel-ironselenium hydrogenases—an overview. *European Journal of Inorganic Chemistry 2011*, no. 7 (March): 948–962. doi:10.1002/ejic.201001127.

Basak, N., D. Das 2007. The prospect of purple non-sulfur (PNS) photosynthetic bacteria for hydrogen production: the present state of the art. *World Journal of Microbiology and Biotechnology 23*, no. 1 (January): 31–42. doi:10.1007/s11274-006-9190-9.

Benemann, J.R. 1997. Feasibility analysis of photobiological hydrogen production. *International Journal of Hydrogen Energy 22*, no. 10/11 (October): 979–987. doi:10.1016/S0360-3199(96)00189-9.

Blankenship, R.E., D.M. Tiede, J. Barber et al. 2011. Comparing photosynthetic and photovoltaic efficiencies and recognizing the potential for improvement. *Science 332*, no. 6031 (May): 805–809. doi:10.1126/science.1200165.

Bolton, J.R., S.J. Strickler, J.S. Connolly. 1985. Limiting and realizable efficiencies of solar photolysis of water. *Nature 316* (August): 495–500. doi:10.1038/316495a0.

Brentner, L.B., J. Peccia, J.B. Zimmerma. 2010. Challenges in Developing Bio-hydrogen as a Sustainable Energy Source: Implications for a Research Agenda, *Environmental Science and Technology 44*, no. 7 (March): 2243–2254. doi 10.1021/es9030613.

Brown, K.A., M.B. Wilker, M. Boehm, G. Dukovic, P.W. King. 2012. Characterization of photochemical processes for H2 production by CdS nanorod-[FeFe]-hydrogenase complexes. *Journal American Chemical Society* 134, 5627–5636. doi:10.1021/ja2116348.

Cavinato, C., A. Giuliano, D. Bolzonella, P. Pavan, F. Cecchi. 2012. Bio-hythane production from food waste by dark fermentation coupled with anaerobic digestion process: A long-term pilot scale experience. *International Journal of Hydrogen Energy, 37* (April): 11549–11555. doi:10.1016/j.ijhydene.2012.03.065.

Cheng, S., B.E. Logan. 2007. Sustainable and efficient biohydrogen production via electrohydrogenesis. *Proceedings of the National Academy of Sciences of the USA 104*, no. 7 (September): 18871–18873. doi:10.1073/pnas.0706379104.

Cracknell, J.A., Vincent, K.A., Armstrong, F.A. 2008. Enzymes as working or inspirational electrocatalysts for fuel cells and electrolysis. *Chemical Reviews 108*, no. 7 (July): 2439–2461. doi:10.1021/cr0680639.

Cuendet, P., K.K. Rao, M. Grätzel, D.O. Hall. 1986. Light induced H_2 evolution in a hydrogenase-TiO_2 particle system by direct electron transfer or via rhodium complexes. *Biochimie 68*, no. 1 (February): 217–221. doi:10.1016/S0300-9084(86)81086-0.

Das, D. 2009. Advances in biohydrogen production processes: An approach towards commercialization. *International Journal of Hydrogen Energy 34*, no. 17 (September): 7349–7357. doi:10.1016/j.ijhydene.2008.12.013.

Das, D., T.N. Veziroglu. 2008. Advances in biological hydrogen production processes. *International Journal of Hydrogen Energy 33*, no. 21 (November): 6046–6057. doi:10.1016/j.ijhydene.2008.07.098.

Das, D., T.N. Veziroglu. 2001. Hydrogen production by biological processes: a survey of literature. *International Journal of Hydrogen Energy 26*, no. 1 (January): 13–28. doi:10.1016/S0360-3199(00)00058-6.

Demirbas, A. 2009. *Biohydrogen for Future Engine Fuel Demands*. London: Springer.

Ditzig, J., H. Liu, B.E. Logan. 2007. Production of hydrogen from domestic wastewater using a bioelectrochemically assisted microbial reactor. *International Journal of Hydrogen Energy 32* (April): 2296–2304. doi:10.1016/j.ijhydene.2007.02.035.

Eroglu, E., A. Melis. 2011. Photobiological hydrogen production: recent advances and state of the art. *Bioresource Technology 102*, no. 18 (September): 8403–8413. doi:10.1016/j.biortech.2011.03.026.

Gaponik, N., Talapin, D.V., Rogach, A.L., Hoppe, K., Shevchenko, E.V., Kornowski, A., Eychmuller, A., Weller, H. 2002. Thiol-capping of CdTe nanocrystals: an alternative to organometallic synthetic routes. *Journal of Physical Chemistry B 106*, no. 29 (June): 7177–7185. doi:10.1021/jp025541k.

Ghirardi, M.L., P.W. King, D.W. Mulder, C. Eckert, A. Dubini, P.C. Maness, and J. Yu. 2014. In D. Zannoni, R. De Philippis (eds.), *Microbial BioEnergy: Hydrogen Production, Advances in Photosynthesis and Respiration* 38, 101–135. doi: 10.1007/978-94-017-8554-9_5.

Gupta, S.K., S. Kumari, K. Karen Reddy, F. Bux. 2013. Trends in biohydrogen production: major challenges and state-of-the-art developments. *Environmental Technology 34,* no. 13/14 (October): 1653–1670. doi:10.1080/09593330.2013.822022.

Gust, D., Moore, T.A., Moore, A.L. 2000. Mimicking photosynthetic solar energy transduction. *Accounts of Chemical Research 34,* no. 1 (November): 40–48. doi:10.1021/ar9801301.

Hallenbeck, P.C., D. Ghosh. 2009. Advances in fermentative biohydrogen production: the way forward? *Trends in Biotechnology 27,* no. 5 (May): 287–297. doi:10.1016/j.tibtech.2009.02.004.

Hambourger, M., Gervaldo, M., Svedruzic, D., King, P.W., Gust, D., Ghirardi, M., Moore, A.L., Moore, T.A. 2008. [FeFe]-hydrogenase-catalyzed H2 production in a photoelectrochemical biofuel cell. *Journal of the American Chemical Society 130,* no. 6 (January): 2015–2022. doi:10.1021/ja077691k.

Ihara, M., Nakamoto, H., Kamachi, T., Okura, I., Maeda, M. 2006a. Photoinduced hydrogen production by direct electron transfer from photosystem I crosslinked with cytochrome c(3) to NiFe-hydrogenase. *Photochemistry and Photobiology 82,* no. 6 (November): 1677–1685. doi:10.1111/j.1751-1097.2006.tb09830.x.

Ihara, M., Nishihara, H., Yoon, K.S., Lenz, O., Friedrich, B., Nakamoto, H., Kojima, K., Honma, D., Kamachi, T., Okura, I. 2006b. Light-driven hydrogen production by a hybrid complex of a NiFe-hydrogenase and the cyanobacterial photosystem I. *Photochemical Photobiology* 82, 676–682. doi:10.1562/2006-01-16-RA-778.

Jacobson, M.Z. 2009. Review of solutions to global warming, air pollution, and energy security. *Energy Environmental Sciences 2009,* no. 2 (December): 148–173. doi:10.1039/B809990C.

Khanna, N., Das, D. 2013. Biohydrogen production by dark fermentation. *WIREs Energy Environment* 2, no. 4 (June): 401–421. doi:10.1002/wene.15.

Krassen, H., Schwarze, A., Friedrich, B., Ataka, K., Lenz, O., Heberle, J. 2009. Photosynthetic hydrogen production by a hybrid complex of photosystem I and [NiFe] -hydrogenase. *ACS Nano 3,* no. 12 (November): 4055–4061. doi:10.1021/nn900748j.

Levin, D.B., L. Pitt, M. Love. 2004. Biohydrogen production: prospects and limitations to practical application. *International Journal of Hydrogen Energy 29,* no. 2 (February): 173–185. doi:10.1016/S0360-3199(03)00094-6.

Lubitz, W., Reijerse, E.J., Messinger, J. 2008. Solar water-splitting into H_2 and O_2: design principles of photosystem II and hydrogenases. *Energy and Environmental Science 2008,* no. 1 (June):15–31. doi:10.1039/B808792J.

Lubner, C.E., Applegate, A.M., Knorzer, P., Ganago, A., Bryant, D.A., Happe, T., Golbeck, J.H. 2011. Solar hydrogen-producing bionanodevice outperforms natural photosynthesis. *Proceeding of the National Academy of Sciences USA 108,* no. 52 (December): 20988–20991. doi:10.1073/pnas.1114660108.

Madden, C., Vaughn, M.D., Diez-Perez, I., Brown, K.A., King, P.W., Gust, D., Moore, A.L., Moore, T.A. 2012. Catalytic turnover of [FeFe] -hydrogenase based on single-molecule imaging. *Journal of the American Chemical Society 134,* no. 3 (September): 1577–1582. doi:10.1021/ja207461t.

Melis, A. 2002. Green alga hydrogen production: progress, challenges and prospects. *International Journal of Hydrogen Energy 27,* no. 11/12 (November-December): 1217–1228. doi:10.1016/S0360-3199(02)00110-6.

Melis, A. 2008. Maximizing light utilization efficiency and hydrogen production in microalgal cultures. U.S. Department of Energy: Washington, DC, 2008:187–190. https://www.hydrogen.energy.gov/pdfs/progress07/ii_h_3_melis.pdf.

Melis, A. 2009. Solar energy conversion efficiencies in photosynthesis: minimizing the chlorophyll antennae to maximize efficiency. *Plant Science 177,* no. 4 (October): 272–280. doi:10.1016/j.plantsci.2009.06.005.

Melis, A., T. Happe. 2001. Hydrogen production green algae as a source of energy. *Plant Physiology 127,* no. 3 (November): 740–748. doi:10.1104/pp.010498.

Melis, A., L. Zhang, M. Forestier, M. L. Ghirardi, M. Seibert. 2000. Sustained photobiological hydrogen gas production upon reversible inactivation of oxygen evolution in the green alga Chlamydomonas reinhardtii. *Plant Physiology 122,* no.1 (January): 127–133. doi:10.1104/pp.122.1.127.

Miura, Y., T. Akano, K. Fukatsu et al. 1997. Stably sustained hydrogen production by biophotolysis in natural day/night cycle. *Energy Conservation Management 38, Supplement*: S533–S537. doi:10.1016/S0196-8904(96)00323-8.

Mussgnug, J.H., S. Thomas-Hall, J. Rupprecht, A. Foo. 2007. Engineering photosynthetic light capture: impacts on improved solar energy to biomass conversion. *Plant Biotechnology Journal* 5, no. 6 (November): 802–814. doi:10.1111/j.1467-7652.2007.00285.x.

Navarro, R.M., M.C. Alvarez-Galvan, J.A. Villoria de la Mano, S.M. Al-Zahrani, J.L.G. Fierro. 2010. A framework for visible-light water splitting. *Energy & Environmental Science 3* (September): 1865–1882. doi:10.1039/C001123A.

Padro, C., V. Putsche. 1999. Survey of the economics of hydrogen technologies. Golden, CO: National Renewable Energy Laboratory, U.S. Department of Energy. https://www1.eere.energy.gov/hydrogenandfuelcells/pdfs/27079.pdf.

Park, H.G., J.K. Holt. 2010. Recent advances in nanoelectrode architecture for photochemical hydrogen production. *Energy & Environmental Science 3* (June): 1028–1036. doi:10.1039/B922057G.

Pedroni, P., G.M. Mura, G. Galli, C. Pratesi, L. Serbolisca, G. Grandi. 1996. The hydrogenase from the hyperthermophilic archaeon *Pyrococcus furiosus*: from basic research to possible future applications. *International Journal of Hydrogen Energy 21,* no. 10 (October): 853–858. doi:10.1016/0360-3199(96)00020-1.

Prince, R.C., H.S. Kheshgi. 2005. The photobiological production of hydrogen: potential efficiency and effectiveness as a renewable fuel. *Critical Reviews in Microbiology 31,* no. 1 (October): 19–31. doi:10.1080/10408410590912961.

Redwood, M.D., M.P. Beedle, L.E. Macaskie. 2009. Integrating dark and light biohydrogen production strategies: towards the hydrogen economy. *Reviews in Environmental Science and Biotechnology 8,* no. 2 (June): 140–185. doi:10.1007/s11157-008-9144-9.

Reisner, E. 2011. Solar hydrogen evolution with hydrogenases: from natural to hybrid systems. *European Journal of Inorganic Chemistry 2011,* no. 7 (December): 1005–1016. doi:10.1002/ejic.201000986.

Reisner, E., D.J. Powell, C. Cavazza, J.C. Fontecilla-Camps, F.A. Armstrong. 2009b. Visible light-driven H_2 production by hydrogenases attached to dye-sensitized. *Journal of American Chemical Society 131,* no. 51 (November): 18457–18466. doi:10.1021/ja907923r.

Reisner, E., J.C. Fontecilla-Camps, F.A. Armstrong. 2009a. Catalytic electrochemistry of a [NiFeSe]- hydrogenase on TiO_2 and demonstration of its suitability for visible-light driven H_2 production. *Chemical Communications 2009,* no. 5 (February): 550–552. doi:10.1039/B817371K.

Rousset, M., L. Cournac. 2008. Towards hydrogenase engineering for hydrogen production. In J. D. Wall, C. S. Harwood, A. Demain (eds.), *Bioenergy,*

ASM Press: Washtington, DC, 249–258. www.asmscience.org/content/book/10.1128/9781555815547.ch20. doi:10.1128/9781555815547.ch20.

Rozendal, R.A., H.V.M. Hamelers, G.J.W. Euverink, S.J. Metz, C.J.N. Buisman. 2006. Principle and perspectives of hydrogen production through biocatalyzed electrolysis. *International Journal Hydrogen Energy 31,* no. 12 (September): 1632–1640. doi:10.1016/j.ijhydene.2005.12.006.

Rupprecht, J., B. Hankamer, J.H. Mussgnug, G. Ananyev, C. Dismukes, O. Kruse. 2006. Perspectives and advances of biological H2 production in *microorganisms. Applied Microbiology Biotechnology 72,* no. 3: 442–449. doi:10.1007/s00253-006-0528-x.

Selvaggi, A., C. Tosi, U. Barberini, E. Franchi, F. Rodriguez, P. Pedroni. 1999. In vitro hydrogen photoproduction using *Pyrococcus furiosus* sulfhydrogenase and TiO_2. *Journal of Photochemistry and Photobiology 125,* no. 1/3 (August): 107–112. doi:10.1016/S1010-6030(99)00088-X.

Show, K. Y., Lee, D.J., Tay, J.H., Lin, C.Y., J. S. Chang. 2012. Biohydrogen production: current perspectives and the way forward. *International Journal of Hydrogen Energy 37,* no. 20 (October): 15616–15631. doi:10.1016/j.ijhydene.2012.04.109.

Shumilin, I.A., V.V. Nikandrov, V.O. Popov, A. A. Krasnovsky. 1992. Photogeneration of NADH under coupled action of CdS semiconductor and hydrogenase from *Alcaligenes eutrophus* without exogenous mediators. *FEBS Letters 306,* no. 2/3 (July): 125–128. doi:10.1016/0014-5793(92)80982-M.

Tekucheva, D.N. A.A. Tsygankov. 2012. Combined biological hydrogen-producing systems: A review. *Applied Biochemistry and Microbiology, 48,* no.4 (July): 319–337. doi:10.1134/S0003683812040114.

Turner, J., G. Sverdrup, M.K. Mann, P.-C. Maness, B. Kroposki, M.L. Ghirardi, R.J. Evans, D. Blake. 2008. Renewable hydrogen production. *International Journal of Energy research 32,* no. 5 (April): 379–407. doi:10.1002/er.1372.

Utschig, L.M., S.C. Silver, K.L. Mulfort, D.M. Tiede. 2011. Nature-driven photochemistry for catalytic solar hydrogen production: a photosystem I transition metal catalyst hybrid. *Journal of the American Chemical Society 133,* no. 41 (September): 16334–16337. doi:10.1021/ja206012r.

Vincent, K.A., J.A. Cracknell, J.R. Clark, M. Ludwig, O. Lenz, B. Friedrich, F.A. Armstrong. 2006. Electricity from low-level H 2 in still air-an ultimate test for an oxygen tolerant hydrogenase. *Chem Commun (Camb) 48* (December): 5033–5035. doi:10.1039/b614272a.

Wang, M., L. Chen, X. Li, L. Sun. 2011. Approaches to efficient molecular catalyst systems for photochemical H 2 production using [FeFe]- hydrogenase active site mimics. *Dalton Trans 40* (October): 12793–12800. doi:10.1039/C1DT11166C.

Wilkes, J., J. Moccia, P. Wilczek, R. Gruet, V. Radvilaite. 2011. Report on EU Energy Policy to 2050. Achieving 80–95% Emissions Reduction. 1–68. www.ewea.org/fileadmin/ewea_documents/documents/publications/reports/EWEA_EU_Energy_Policy_to_2050.pdf.

Zabrosky, O.R., editor. 1997. *Biohydrogen. Proceeding of International Conference on Biological Hydrogen Production*. Waikoloa, Hawaii: Plenum Press.

Zhang, J.D., A.C. Welinder, Q.J. Chi, J. Ulstrup. 2011. Electrochemically controlled self-assembled monolayers characterized with molecular and sub-molecular resolution. *Physical Chemistry Chemical Physics 2011,* no. 13 (February): 5526–5545. doi:10.1039/C0CP02183K.

7

Photocatalytic CO_2 Reduction to Fuels

Biswajit Mishra and Yatendra S. Chaudhary

CONTENTS

7.1 Introduction

Photocatalytic reduction of CO_2 to energy-rich fuels such as methanol or methane is not only lucrative in carbon neutral energy perspective but also carries a lot of significance in mitigating the ever-growing anthropogenic CO_2 level in atmosphere. Various approaches and intrigued prospects related to the solar H_2 generation have already been detailed in previous chapters. However, hydrogen being the lightest element, transportation and storage of H_2, after its generation from solar water splitting, are challenging and require special care and complete overhaul of the existing storage and transportation set-up. On the contrary, energy-rich liquid or gas, such as methanol, ethanol, or methane, obtained from the photocatalytic reduction of CO_2 can easily be stored and transported using the existing infrastructure that is already proficient in transportation of natural gas and liquid fuels (Morris et al. 2009). Particularly this makes those CO_2 reduction products even more attractive. Also some of the seemingly unimportant products obtained on CO_2 photoconversion can suitably be used for various chemical syntheses. Therefore, CO_2 may become a potential carbon neutral energy source and an alternative for fast depleting fossil fuels soon (Habisreutinger et al. 2013). Much attention has already been paid in research of the photocatalytic CO_2 reduction since the first report of photoreduction of CO_2 using several semiconductor photocatalysts by Inoue et al. (1979). However, numerous challenges are preventing this domain of research to flourish from its infancy. Still the highest reported rates of product formation are very low and are in the range of only few tens of μmoles per hour of illumination using 1 g of photocatalyst. The exact mechanism behind the photoreduction of CO_2 follows very complex paths and is yet to be understood clearly. The two major challenges retarding the progress are selectivity and energetics. These are elaborated in the succeeding sections.

7.2 Challenges in CO_2 Photoreduction: Energetics and Selectivity

CO_2 is among the most stable molecules (ΔG_f^0 = −394.4 kJ/mole) and possesses the highest of all probable oxidation states (−4 to + 4) for carbon. Photocatalytic CO_2 reduction involves proton-assisted transfer of multiple electrons (Equations 7.2 to 7.6). Therefore, it is quite explicable that the CO_2 reduction requires high energy for bond breaking (749 kJ/mole of C=O bond) and often leads to formations of various products with different oxidation states of the carbon. In addition, hydrogen evolution reaction (HER) also takes place (Equations 7.7 and 7.8) in competition with the photocatalytic reduction of CO_2 in the presence of water (vapor or liquid phase) (Chang et al. 2015; Habisreutinger et al. 2013). Therefore, the energetics and selectivity remain major challenges in photocatalytic reduction of CO_2. The underlying principle, as discussed in Sections 2.3 and 3.1 for the photocatalysis process, is that the photogenerated excitons (electron and hole pairs) should have sufficient redox potential to drive the fuel forming reactions (Equations 7.1 through 7.8) at the surface of the photocatalyst. Before going into the detailed discussion about the selectivity and the energetics related to the photoreduction of CO_2, we should take a look at the redox potentials required to drive important reactions during the course of CO_2 photoreduction. Also Scheme 7.1 represents the positions of conduction band minimum E_c and valence band maximum E_v of various semiconductor photocatalysts with respect to the potentials of the redox reactions mentioned below (Habisreutinger et al. 2013).

$$CO_2 + e^- = CO_2^- \qquad E^0_{redox} = -1.9 \text{ V vs NHE at pH 7} \tag{7.1}$$

$$CO_2 + 2H^+ + 2e^- = HCOOH \qquad E^0_{redox} = -0.61 \text{ V vs NHE at pH 7} \tag{7.2}$$

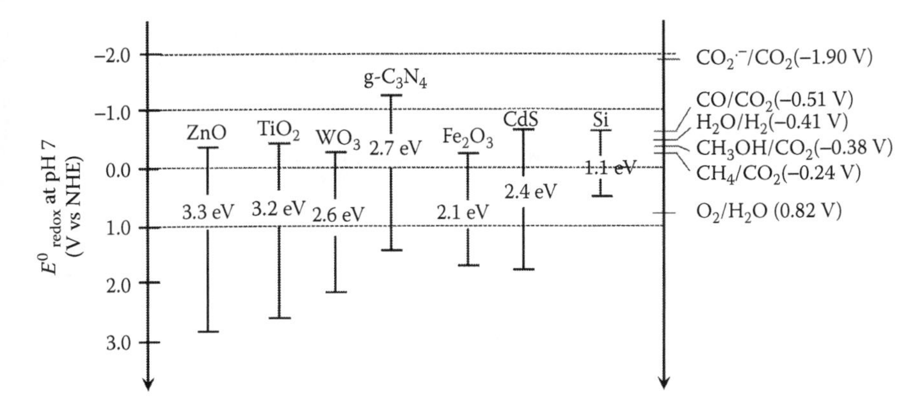

SCHEME 7.1
Band gaps, positions of E_c and E_v of various popular semiconductor photocatalysts, and the potentials of the redox couples participating in fuel generation reactions (Equations 7.1 through 7.8).

$$CO_2 + 2H^+ + 2e^- = CO + H_2O \qquad E^0_{redox} = -0.51 \text{ V vs NHE at pH 7} \tag{7.3}$$

$$CO_2 + 4H^+ + 4e^- = HCHO + H_2O \qquad E^0_{redox} = -0.48 \text{ V vs NHE at pH 7} \tag{7.4}$$

$$CO_2 + 6H^+ + 6e^- = CH_3OH + H_2O \qquad E^0_{redox} = -0.38 \text{ V vs NHE at pH 7} \tag{7.5}$$

$$CO_2 + 8H^+ + 8e^- = CH_4 + H_2O \qquad E^0_{redox} = -0.24 \text{ V vs NHE at pH 7} \tag{7.6}$$

$$2H^+ + 2e^- = H_2 \qquad E^0_{redox} = -0.41 \text{ V vs NHE at pH 7} \tag{7.7}$$

$$2H_2O = O_2 + 4H^+ + 4e^- \qquad E^0_{redox} = 0.82 \text{ V vs NHE at pH 7} \tag{7.8}$$

It is quite evident from Scheme 7.1 that the proton-assisted transfer of multiple electrons (Equations 7.2 through 7.6) is thermodynamically feasible as the E_c of several semiconductors is more negative as compared to that of potential of those redox reactions (Equations 7.2 through 7.8). However, there is no such report in the literature in support of the proton-assisted transfer of multiple electrons in a single step. Rather it is widely perceived that the photoreduction of CO_2 proceeds via several one-electron steps sequentially initiated by the formation of $CO_2^{\cdot-}$ through one electron transfer from conduction band (CB) of semiconductor photocatalyst to the lowest unoccupied molecular orbital (LUMO) of CO_2 (Equation 7.1). A very high negative potential (−1.9 V vs NHE at pH 7) is required to trigger the first step and hardly any semiconductor is available to drive this first step (Scheme 7.1). Hence this step turns out to be the bottleneck for CO_2 photoreduction (Habisreutinger et al. 2013). Numerous excellent theoretical and experimental studies nicely explained how different adsorption and binding modes of CO_2 on semiconductor photocatalysts' surface play crucial roles not only in activating CO_2 molecules to trigger the thermodynamically unfavorable one electron transfer from CB of semiconductor photocatalyst to the LUMO of CO_2, but also help in guiding the overall reaction toward the formation of a specific product (Chang et al. 2015; Gattrell et al. 2007; Tu et al. 2014). These phenomena address two major challenges: energetics and selectivity. These are detailed in the next section.

7.3 Photocatalytic CO_2 Reduction

7.3.1 Semiconducting Photocatalysts

Semiconductor-based heterogeneous catalysis has been studied extensively since 1979 (Inoue et al. 1979) and covers the major part of research studies carried out in the esteemed field of CO_2 photoreduction. The use of semiconductors not only harvests light to generate excitons but also provides a platform for the adsorption and activation of CO_2 on their surfaces. Therefore, to understand the mechanism behind CO_2 adsorption and activation is inevitable to devise an effective photocatalyst for CO_2 reduction. As mentioned in the previous section, the activation of chemically stable CO_2, which involves one electron transfer from excited photocatalysts to the LUMO of CO_2 to generate $CO_2^{\cdot-}$ species, is thermodynamically unfavorable. However, CO_2 loses its linearity on binding on the surface of a semiconductor through either donation of lone pairs of electrons residing on its two oxygen atoms to surface Lewis acid centers (Figure 7.1a), carbon gaining electron

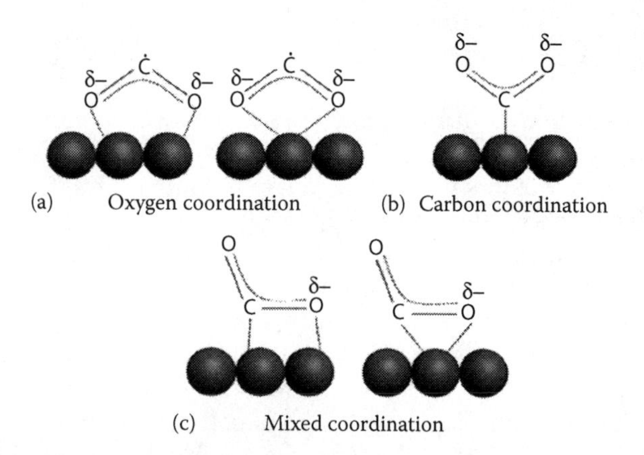

FIGURE 7.1
The possible structures of adsorbed $CO_2\delta^{\bullet-}$ on catalysts. (Data from Gattrell, M. et al., *Journal of Electroanalytical Chemistry*, 594, 1–19, 2007; Freund, H. J., and R. P. Messmer, *Surface Science*, 172, 1–30, 1986; Freund, H. J., and M. W. Roberts, *Surface Science Reports*, 25, 227–273, 1996; Chang, X. et al., *Energy & Environmental Science*, Advanced Article, 2016; Reproduced from Chang et al. 2016 with permission from the Royal Society of Chemistry.)

from surface Lewis base centers like oxide ions (Figure 7.1b), or the combination of both (Figure 7.1c) (Chang et al. 2015; Gattrell et al. 2007; Tu et al. 2014). The LUMO of CO_2 lowers in energy with the decrease in the O–C–O bond angle. Consequently, activation of CO_2 through the transfer of one electron from semiconductor to CO_2 LUMO becomes easier after its binding on the semiconductor surface. Depending on the extent of interactions through the binding of CO_2 on the semiconductor surface, the deviation of the O–C–O bond angle from 180° varies and hence the LUMO energy of CO_2 does so (Figure 7.2) (Freund and Roberts 1996; Indrakanti et al. 2009). Therefore, it is evident that the strong binding of CO_2 on the semiconductor surface is essential for the activation of CO_2 to be thermodynamically feasible. This holds true even in the case of the CO_2 activation on the electrocatalyst to decrease the overpotential for CO_2 reduction (Schmidt et al. 1994). There are several strategies that have been reported to enhance the chemisorption of CO_2 on semiconductors' surface and improve the rate of specific product formation, which will be categorically discussed in the rest of this section.

7.3.1.1 Creating Surface Defects

Defects such as oxygen vacancy (Liu et al. 2012), oxygen rich (Tan et al. 2014), or sulfur vacancy (Fujiwara et al. 1997; Kanemoto et al. 1996) on the surface of a semiconductor play crucial roles in enhancing chemisorption of CO_2 providing specific binding sites and hence enhanced product formations and selectivity. For example, creating controlled oxygen vacancy on the surface of TiO_2 (or partially reduced TiO_2) not only helps in the formation of a specific product as CO from the CO_2 photoreduction but also increases the rate

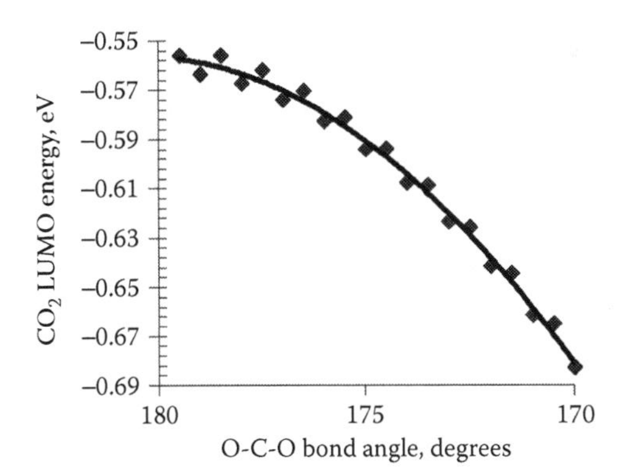

FIGURE 7.2
The variation in the energy of the LUMO of gaseous CO_2 with the O–C–O bond angle, calculated using constrained quantum chemical calculations at the B3LYP/6-31 + G(d) level of theory. Decreasing the O–C–O bond angle (via surface interactions) may facilitate charge transfer to CO_2 by lowering its LUMO. (Data from Freund, H. J., and M. W. Roberts, *Surface Science Reports*, 25, 227–273, 1996; Indrakanti, V. P. et al., *Energy & Environmental Science*, 2, 745–758, 2009. Reproduced by permission of the Royal Society of Chemistry.)

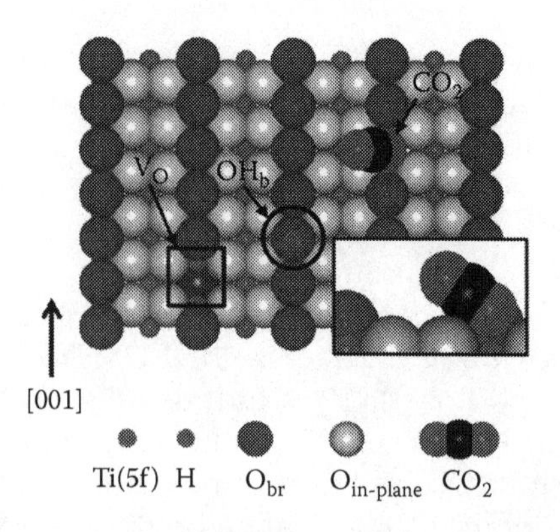

FIGURE 7.3
Schematic showing an oxygen vacancy defect (V_O) (black square), a bridging hydroxyl (OH_b) (black circle), and a CO_2 molecule adsorbed at a V_O site on the reduced TiO_2(110)-(1 × 1) surface. Five-fold-coordinated Ti(5f) atoms and bridging oxygens (O_{br}) are indicated as smallest and largest dark gray spheres, respectively. The molecular axis of the adsorbed CO_2 is perpendicular to the direction of the bridging-oxygen row ([001] azimuth) and is tilted away from the surface normal by 57°, as shown in the inset figure. (Reprinted with permission from Lee, J. et al., *Journal of American Chemical Society*, 133, 10066–10069, Copyright [2011] the American Chemical Society.)

of CO formation dramatically as compared to fully oxidized TiO_2. One of the two oxygen atoms of CO_2 binds at the oxygen vacancy sites (V_O in Figure 7.3) strongly (Lee et al. 2011). This facilitates in activating CO_2 and the removal of CO leaving behind one oxygen atom at the V_O site for healing the defect. Photoconversion efficiency largely depends on the density and distribution of the defects (Kong et al. 2011). Consequently, the processes followed in creating oxygen vacancies are also crucial. Oxygen vacancies can be created through annealing under vacuum (Dillip et al. 2016) or under the flow of inert gases such as He or Ne (Liu et al. 2012). These processes are both energy and time consuming, and may affect the crystallinity and the morphology of the nanostructures. However, the flame reduction method at a temperature as high as 1000°C offers unique advantages such as ultrafast heating rate, tunable reduction environment, and open-atmosphere operation. So oxygen vacancies can be created in less than 1 min without deteriorating the crystallinity and the morphology of the nanostructures (Cho et al. 2014). Contrary to the oxygen vacancy, as a unique example, the oxygen-rich surface of anatase TiO_2 (O_2-TiO_2) drives CO_2 photoreduction toward the formation of CH_4 in a much faster rate (5.7 times) as compared to pure anatase TiO_2. However, oxygen attachment on the TiO_2 surface, which can be prepared in a very simple way by just modifying TiO_2 precursor with H_2O_2, helps in reducing charge carrier recombination (Etacheri et al. 2011).

There are other examples of similar enhancements in product formations and selectivity when sulfur vacancies are created on the surface of sulfide-based photocatalysts. This can be done, for example, in the case of CdS, by adding excess of Cd^{2+} in *N,N*-dimethylformamide. This results in the formation of sulfur vacancies, *in situ*, on the surface of nanocrystallites due to the adsorption of excess Cd^{2+} to the surface and a remarkable increase of photocatalytic activity for CO_2 reduction to CO (Fujiwara et al. 1997). Theoretical calculation shows enhanced adsorption of CO_2 to the Cd atom in the vicinity of sulfur vacancy.

7.3.1.2 Nanostructuring of Semiconductor Catalysts

Selective exposure of a specific plane also greatly influences the rate and selectivity in product formations. The anatase phase of TiO_2, the most photoactive among the three phases: anatase, rutile, and brookite, contains the thermodynamically most stable (101) planes (surface energy: 0.44 J/m^2) at the surface. However, the (001) and (100) planes (surface energy: 0.90 J/m^2 and 0.53 J/m^2, respectively) (Gong and Selloni 2005; Lazzeri et al. 2001, 2002; Wen et al. 2011) are known to be the most reactive due to the presence of dangling bonds from Ti and O atoms (Han et al. 2009). When the (101) and (001) planes are coexposed in the optimal ratio (Figure 7.4a), the photocatalytic activity toward the reduction of CO_2 to CH_4 is significantly enhanced (Yu et al. 2014). This is mainly due to the formation of a surface heterojunction within a single TiO_2 particle. As shown in Figure 7.4b, photogenerated electrons migrate to the (101) plane and holes to the (001) plane. Consequently, this reduces charge carrier recombination.

7.3.1.3 Surface Modification with Basic Sites

Functionalizing the surface of semiconductors with basic sites such as MgO, NaOH, or ammines is also another effective approach to improve trapping and activation of CO_2 on the semiconductor surface. Loading an

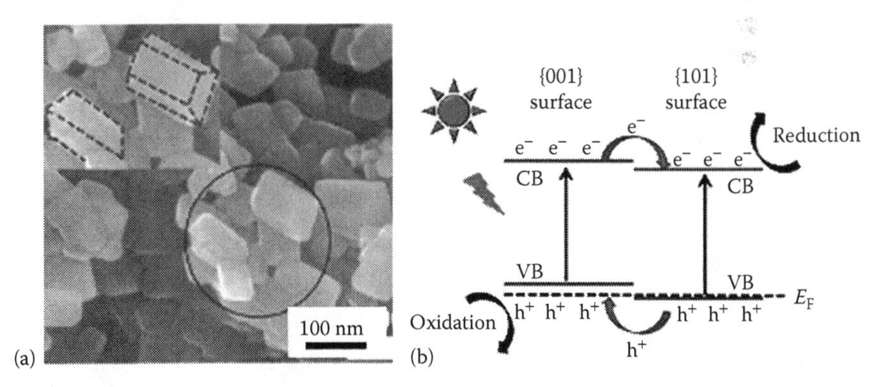

FIGURE 7.4
(a) FESEM images of anatase with ratio of 45:55 coexposed {101} and {001} facets and the enlarged view of the image in the inset and (b) {001} and {101} surface heterojunction. (Reprinted with permission from Yu, J. et al., *Journal of American Chemical Society*, 136, 8839–8842, 2014. Copyright [2014] the American Chemical Society.)

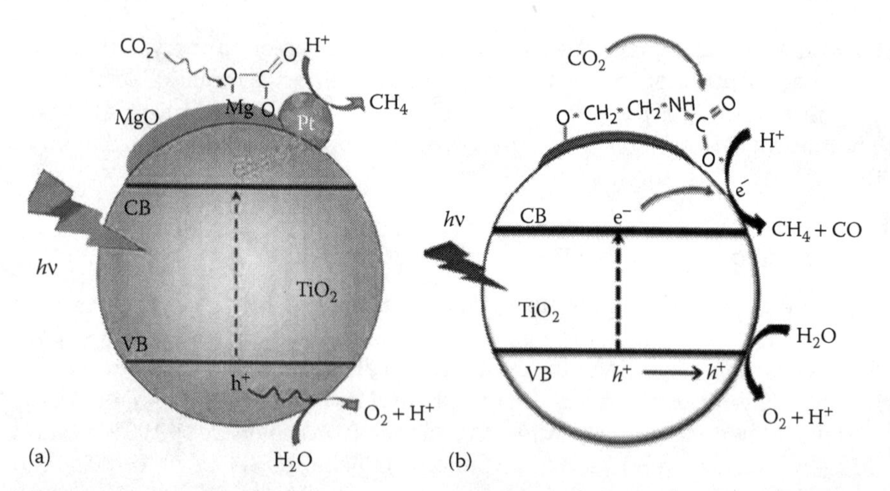

FIGURE 7.5
(a) Proposed functioning mechanisms of the MgO layer and Pt nanoparticles over TiO_2 for the photocatalytic reduction of CO_2 in the presence of H_2O. (Xie, S. et al., *Chemical Communications*, 49, 2451–2453, 2013. Reproduced by permission of the Royal Society of Chemistry.) (b) Proposed mechanism of MEA–TiO_2 capture of CO_2 on the TiO_2 surface and the photochemical reduction to CO and CH_4. VB = valence band, CB = conducting band. (Reprinted with permission from Liao, Y. et al., Efficient CO_2 capture and photoreduction by amine-functionalized TiO_2, *Chemistry A European Journal*, 2014, 20, 10220–10222. Copyright Wiley-VCH Verlag GmbH & Co. KGaA. Reproduced with permission.)

optimal amount of MgO on Pt-TiO_2 considerably enhances the photocatalytic reduction of CO_2 with H_2O to CH_4 predominantly. The interface in this trinary system of TiO_2, Pt, and MgO is very important in the photocatalytic reaction. The presence of MgO on the TiO_2 surface helps better in CO_2 adsorption and activation, whereas nearby Pt nanoparticles act as the electron sink and pass the photogenerated electrons from TiO_2 to CO_2 molecules (Figure 7.5a) (Xie et al. 2013). Similarly, surface modification of TiO_2 with NaOH shows better CO_2 adsorption, activation, and finally effective conversion of CO_2 into CH_4 (Meng et al. 2014). However, significant amount of H_2 also evolves along with CH_4. In case of functionalization of TiO_2 nanoparticles with ammine through a solvothermal approach, the mixture of CO and CH_4 evolves. Here ammine functionalization substantially increases the extent of CO_2 chemisorption on TiO_2 surfaces and consequently more effective activation, charge transfer from excited TiO_2 to CO_2, and finally reduction into methane and CO (Figure 7.5b) (Liao et al. 2014). This ammine-modified TiO_2 shows significant amount of CO_2 uptake even at relatively low CO_2 pressure levels, which suggests that specific interactions occur between CO_2 and ammine-modified TiO_2 during the CO_2 adsorption process.

7.3.1.4 Formation of Heterostructures

All of the surface modifications mentioned above improve the rate of product formation significantly as compared to that of the unmodified semiconductors through enhancing the surface adsorption and activation of CO_2. The formation of heterostructure photocatalysts with appropriate energetics has an additional advantage, as it improves the exciton separation. Depending on the alignments of the band edges of the two constituent semiconductors, heterostructures may contain three types of heterojunctions: types I, II, and III, which has been discussed in Section 4.2 in detail. Among them, type II turns out to be the most beneficial, and its formation will only be discussed here (Scheme 7.2a). Precisely, in type II heterostructures, photogenerated electrons and holes move away from each other across the junctions (specifically Schottky (p–n) junction). These separated excitons now reside at CB and valence band (VB) respectively of the two different semiconductors (SC2 and SC1). Consequently, less recombination occurs now as they are spatially separated from each other, and this leads to improved photocatalytic activity. There are few heterostructures reported in the literature to drive photoreduction of CO_2. Among them, in recent years, heterostructures containing $TiO_2/g\text{-}C_3N_4$ have gained popularity as the potential candidate for CO_2 photoreduction owing to their perfect band edge matching. However, individually both TiO_2 and $g\text{-}C_3N_4$ are well known as photocatalysts due to their own merits in terms of low cost, long-term stability, easy availability, and nontoxicity, and these make them ideal candidates to be coupled. *In situ* grown composite between $g\text{-}C_3N_4$ (SC1) and N-TiO_2 (SC2) with the optimum ratio by a facile heating method indeed demonstrates much better photocatalytic performance in photoreduction of CO_2 in the presence of water vapor not only in terms of product yield but also in terms of product selectivity (Zhou et al. 2014). The composite produces almost exclusively CO only with the highest evolution amount of 14.73 μmol under 12 h of light irradiation, whereas in the case of a N-TiO_2 sample, both CO and CH_4 evolve. As it is already reported by Lin et al. (2014) that the CO_2 adsorption and activation can be enhanced on $g\text{-}C_3N_4$, this betterment in product yield and product selectivity can be correlated to the synergy between better separation of the photogenerated charge carriers and enhanced chemisorption of CO_2 on $g\text{-}C_3N_4$. Therefore, this result serves as an inspiration to design and develop heterostructured photocatalysts to drive photoreduction of CO_2 efficiently.

7.3.1.5 Z-Scheme for Photocatalytic CO_2 Reduction

It is clear from the above discussion that the formation of heterostructures, though, enhances light harvesting and charge separation; one can realize that the preference in actual functional property (preference of SC2 in water oxidation due to having more positive E_v and SC1 in CO_2/H^+ reduction due

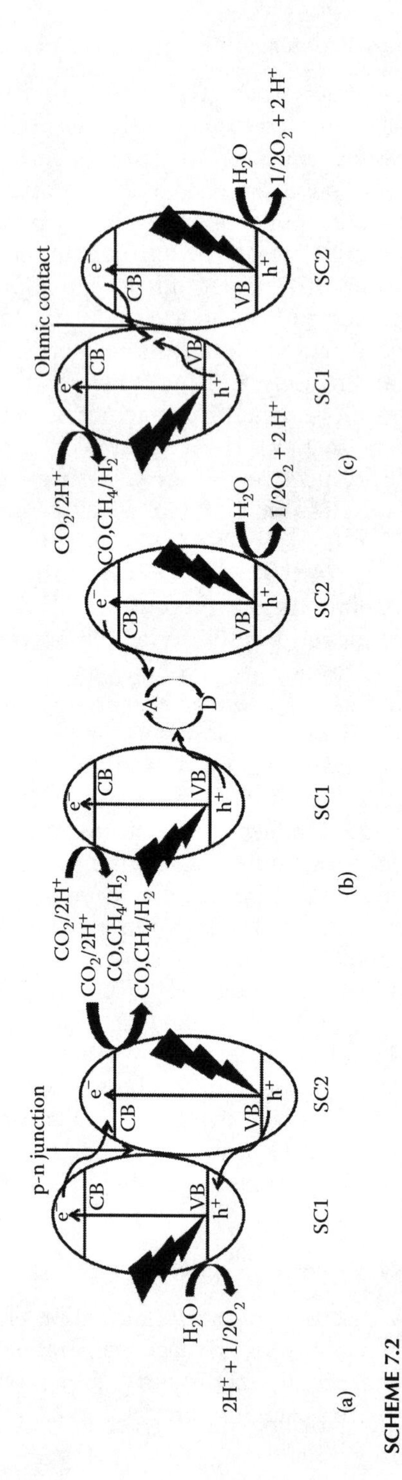

SCHEME 7.2
Schematics for (a) heterostructure with type II junction, (b) artificial Z-scheme with liquid electron mediator, and (c) all solid-state Z-scheme heterostructured photocatalysis systems.

to having more negative E_c) of the semiconductor reverses after forming the type II junction, and both electrons and holes lose their redox capabilities to some extent after their separation. In the extreme case of energy offset, the conjugated system may even no longer remain active either of water oxidation and CO_2/H^+ reduction or both. Rather scientific community, inspired by the natural photosynthesis system (discussed in Section 1.2, Figure 1.3), has been focusing on combining SC1 and SC2 using an electron mediator between photogenerated electrons in CB of SC2 and holes in VB of SC1. These conjugated systems, popularly known as the artificial Z-scheme (Scheme 7.2b), are gaining much interest even in CO_2 reduction mainly photoelectrochemically and also photocatalytically. Unlike type II, in the Z-scheme, electrons and holes with weaker redox potentials recombine with each other keeping electrons and holes with stronger redox potentials available for the CO_2/H^+ reduction and water oxidation, respectively. However, this artificial Z-scheme photocatalytic system is also associated with some disadvantages. The electron mediators used in these systems take part in backward reactions also, that is, mediate between the holes and electrons from the VB of SC2 and CB of SC1, respectively, which retard the rate of photocatalysis. Also their activity and stability largely depend on a precise pH range in solution and hence any deviation from that range will affect the stability and the available numbers of donor acceptor pairs largely. Most importantly, these systems can only be applicable for the photocatalytic reactions in the liquid phase only, whereas photocatalytic CO_2 reductions are preferably performed in the gas phase in the presence of CO_2 and water vapor (Fan et al. 2013; Qu et al. 2013; Zhou et al. 2014). In contrast to the above-mentioned artificial Z-scheme photocatalytic system, the design of a less explored all solid-based Z-scheme photocatalytic system without any electron mediators is more beneficial. SC1 are carefully deposited on the surface of SC2 in such a way that their interface will act as an ohmic contact with minimum contact resistance and also as a mediator between holes and electrons from the VB of SC1 and CB of SC2, respectively. The factors controlling the formation of the all solid-state Z-scheme have been elaborated in Section 4.3. Recently there have been several reports of the Z-scheme (both of using electron mediators and all solid state), however, only for hydrogen production and degradation reactions concerning several pollutants such as formaldehyde (Yu et al. 2013), phenol (Liao et al. 2015), toluene (Munoz-Batista et al. 2014), where it has been shown that the heterostructures behaved actually in the all solid Z-scheme manner. Reports for the photoreduction of CO_2 are scarce. Most of the CO_2 photoreduction were done electrochemically (Arai et al. 2010; Sato et al. 2010, 2011) and few are photocatalytically (Li et al. 2015; Sekizawa et al. 2012; Wang et al. 2015, 2016).

Sekizawa et al. (2012), for the first time, presented a hybrid Z-scheme photocatalyst driven by exclusively visible light. The hybrid system was formed combining the Ru(II) dinuclear complex (RuBLRu′) on Ag-loaded TaON (Ag/TaON). When this system was applied in photocatalytic CO_2 reduction under the illumination of visible light using methanol as a sacrificial reducing agent, mainly formic acid was formed with turnover numbers (TONs) 41 in 9 h. The additional advantage of this system is that, unlike the other systems containing molecular catalysts, this system does not require stringent electron donors such as triethylamine (TEA) or triethanolamine (TEOA).

A hybrid Z-scheme photocatalyst for CO_2 photoreduction containing rather inexpensive materials: g-C_3N_4 and Bi_2WO_6 was reported (Li et al. 2015). Hydrothermally, *in situ* grown composites showed greatly improved light harvesting, and subsequent improvement in product selectivity for the photoreduction of CO_2 to CO. The optimum rate for CO production was 5.19 μmol/g h, which was higher by 22 and 6.4 times as compared to that obtained with bare g-C_3N_4 and Bi_2WO_6, respectively.

A nanocomposite BiOI/g-C_3N_4 with the optimum molar ratio of 7.4 forms a Z-scheme photocatalyst. It showed the optimum yields for CO, CH_4, and H_2 to be 4.86, 0.18, and 2.78 μmol/g_{cat} h, respectively, after 3 h of visible light illumination (Wang et al. 2016). Therefore, these results appear promising in designing new types of Z-scheme photocatalysts for practical applications.

7.3.1.6 Surface Modification with Cocatalysts

It is well known that assembling noble metal cocatalysts (i.e., Pt, Pd, Au, Ag) onto the surface of semiconductor photocatalyst accelerates charge separation and reduces the overpotential required for the water redox reaction. Noble metals serve as electron sink and provide effective proton reduction sites and hence dramatically facilitate photocatalytic H^+/CO_2 reduction (Yang et al. 2013). As noble metals are rare and costly, there is a need of alternate cocatalyst consisting of earth abundant materials, and with comparable performance. To address this issue, a heterogeneous photocatalyst system consisting of a ruthenium complex anchored to carbon nitride (C_3N_4), which facilitates in photocatalysis by simultaneously providing a catalytic site and better light harvesting, was developed for the reduction of CO_2 into formic acid. The immobilized metal complex accepts an excited electron from the conduction band of C_3N_4 and hosts the active site for CO_2 reduction. The larger potential difference between the CB of C_3N_4 and redox potential of RuP drives electrons faster and inhibits the back electron transfer. The maximum photocatalytic activity obtained for the conversion of CO_2 into HCOOH with TONs as high as 1000 and quantum yield Φ as 5.7% at 400 nm (Kuriki et al. 2015).

In another study, rather inexpensive and more abundant cocatalyst (cobalt species: molecular as well as nanoparticles) was combined with already cheap and stable g-C_3N_4. g-C_3N_4, as mentioned in Section 7.3.1.3, can provide adsorption and activation sites for CO_2. In addition, it also acts as a photosensitizer to the cocatalyst on its surface. The molecular cobalt species facilitate in CO_2 reductions, whereas CoO_x nanoparticles mediate holes, with the electron donors present in the solution, after extracting holes from g-C_3N_4. The maximum yield was obtained predominantly for CO of 3.7 μmol in 2 h with $\Phi = 0.25\%$ at 420 nm. Apart from that, this hybrid photocatalyst system is complete devoid of any expensive metal and shows excellent recyclability. It shows promises in moving forward for artificial CO_2 fixing with cheap and all-sustainable systems.

7.3.2 Molecular Photocatalysts

Transition metal-based molecular catalysts also have their own merits in the photocatalytic reduction of CO_2. Their reactivity, redox potential, or light harvesting capability can easily be tuned through simple synthetic modifications at a molecular level. Also their product selectivity and overall quantum efficiencies are better as compared to that of semiconducting photocatalysts. In general, molecular photocatalysts broadly consist of three parts: a photosensitizer (P), a donor (D), and a catalyst (Cat). Ever since the first report by Lehn et al. on photocatalytic CO_2 reduction using a molecular catalyst (Lehn and Ziessel 1982) ruthenium(II) trisbipyridine, $[Ru(bpy)_3]^{2+}$, has been used as a photosensitizer (P) which, on absorbing light, is promoted to an excited state (P*) (Equation 7.9). TEOA or TEA acts as an electron donor (D) to reductively quench P* to a reduced sensitizer (P^-) and itself is oxidized to $D^{\bullet+}$, here $Et_3N^{\bullet+}$ (Equation 7.10). It is already mentioned that the reduction of CO_2 to CO or HCOOH follows proton-assisted double electron transfer (Equations 7.2 and 7.3). However, the molecular catalysts used in photocatalytic CO_2 reductions are incapable of extracting H^+ from H_2O. Instead reactive $D^{\bullet+}$ assists in proton abstraction and rearranges to $Et_2NC^{\bullet}HCH_3$ (Equation 7.13). In next steps, P^- reverts back to its initial form (P) transferring electron to catalysts (cat) to form the reduced active state, cat^- (Equation 7.11). The cat^- binds to CO_2 and eventually reduces it to CO or HCOOH (Equation 7.12) (Morris et al. 2009; Reithmeier et al. 2012).

$$P + h\nu \rightarrow P^* \tag{7.9}$$

$$P^* + Et_3N \rightarrow P^- + Et_3N^{\bullet+} \tag{7.10}$$

$$P^- + cat \rightarrow P + cat^- \tag{7.11}$$

$$cat^- + CO_2 \rightarrow\rightarrow cat + products \tag{7.12}$$

$$Et_3N^{\bullet+} + Et_3N \rightarrow Et_3NH^+ + Et_2NC^{\bullet}HCH_3 \tag{7.13}$$

In few cases catalysts play a dual role of light harvesting and the catalysis; however, the overall mechanism remains pretty much the same. Before discussing few important outcomes in recent years, which will help to chalk down future prospects in photocatalytic CO_2 reduction, few common terms (Morris et al. 2009) need to be defined as follows:

1. **Catalytic selectivity (CS):** In the case of photocatalytic CO_2 reduction to CO, CH_4, and so on, as mentioned earlier, HER also occurs most of the time in parallel. The CS indicates the molar ratio of the product(s) obtained from CO_2 reduction relative to that of hydrogen produced.

$$CS = [\text{Products from } CO_2 \text{ reduction}]/[\text{Hydrogen}] \quad (7.14)$$

2. **Quantum yield (Φ):** The photochemical quantum yield, φ, can be defined as the molar ratio of products from CO_2 reduction and incident photons.

$$\Phi = [\text{Products from } CO_2 \text{ reduction}]/[\text{incident photons}] \quad (7.15)$$

3. **Turnover numbers (TONs):** The TON can be defined as the number of CO_2 molecules reduced per molecule of catalyst over the catalyst's lifetime. However, in practice, it is often reported for the catalysis time only, irrespective of the fact whether the catalyst is still active or not. This is generally estimated as the molar ratio between the products obtained from CO_2 photoreduction to the catalyst introduced at the beginning.

$$TON = [\text{Products from } CO_2 \text{ reduction}]/[\text{Catalyst}] \quad (7.16)$$

Initially most of the catalysts (cat) reported contain expensive and rare metals such as rhenium (Agarwal et al. 2012; Hawecker et al. 1983) or ruthenium (Tamaki et al. 2012; White et al. 2014). Consequently, it has been difficult to scale up this method of CO_2 reduction for practical purpose. However, in recent years, trends have been shifted toward the formation of more economically viable systems employing inexpensive and abundant metals such as manganese (Bourrez et al. 2011), cobalt (Jeletic et al. 2013), nickel (Thoi et al. 2013), or iron. Saveant group took up the challenge to develop an inexpensive electrocatalyst based on iron(II) porphyrin (FeTPP), which has been already known to be active in CO_2 photoreduction electrocatalytically. However, both its efficiency and the stability were very low. Saveant group gradually improved the proficiency of this inexpensive electrocatalyst (Bhugun et al. 1994). In the first step, in their seminal work in 1994, they have shown that the addition of Lewis acids (specifically Mg^{2+}) to the reaction system dramatically improves the stability of the catalyst, selectivity toward the formation of a specific product (CO, Faradaic efficiency ca. 70%), and obviously the turnover number. The addition of Bronsted acids (e.g., trifluoroethanol) shows similar results as well. This result inspired them to recognize the need of a huge amount of proton donors to be present in the vicinity of the active site (Bhugun et al. 1994, 1996). This drives them to modify FeTPP (structure in Figure 7.6) with phenolic groups in all ortho and ortho′ positions of the phenyl groups in tetraphenyl porphyrin (TPP). After this modification, FeTDHPP (structure in Figure 7.6) shows remarkable improvements in faradaic yield of CO as high as above 90%, 50 million TONs at a low overpotential of 0.465 V and no degradation was observed over 4 h of electrocatalysis (Costentin et al. 2012). Eventually, Routier et al. promoted this electrochemical field through their series of pioneering works

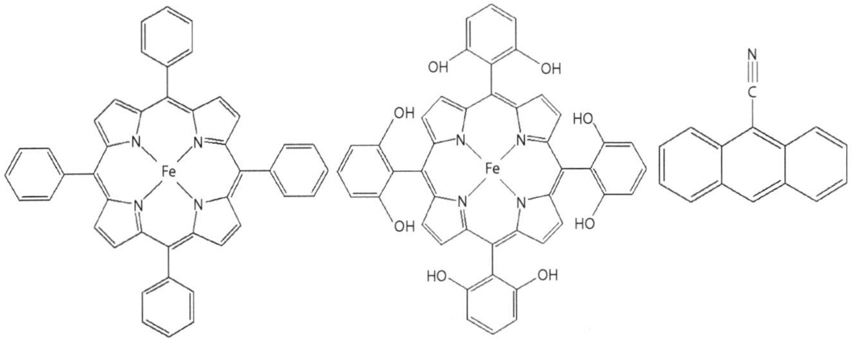

Iron tetraphenyl porphyrin **FeTPP** Iron 5, 10, 15, 20-tetrakis(2′,6′-dihydroxyphenyl) porphyrin **FeTDHPP** 9-Cyanoanthracene **9CNA**

FIGURE 7.6
Structures for different iron porphyrins and 9-cyanoanthracene.

to easily accessible photocatalytic CO_2 reduction (Bonin et al. 2014), where FeTDHPP shows outstanding activity in photocatalytic CO_2 reduction in the presence of a cheaper organic photosensitizer (9-cyanoanthracene, 9CNA, structure in Figure 7.6). It shows CS for CO of 100% and TON for CO ca. 40 under continuous visible light illumination of 45 h; even after that, there was no sign of any degradation of the catalytic system. Therefore, these photocatalytic systems, based on very simple aspects such as an abundant metal (iron)-based catalyst, a cheap organic sensitizer (9CNA), and visible photons, carry promise to be able to raise this field to the next level of devising a basic technological device.

7.3.3 Biomimetic/Enzyme-Based Photocatalysts

Biomimetic or enzyme-based photocatalysis, which is another fascinating way to reduce CO_2 photocatalytically, has gained interest in recent times (Chaudhary et al. 2012; Grodkowski et al. 2000; Woolerton et al. 2010). Grodkowski et al. reported that cobalt corrins, the central structure of vitamin B12, can reduce CO_2 to CO and formic acid as a homogeneous photocatalyst in acetonitrile/methanol solutions where p-terphenyl and triethylamine were used as a photosensitizer and as an electron donor, respectively. Similarly, photoactivity was also measured using the photocatalysts containing porphyrin analogues and compared with the results associated with corrins. The results show better photoactivity and stability of corrins. Maximum selectivity of CO over H_2 obtained for corrins was 53% as compared to that of porphyrins (31%). The total yield of products was four times higher in the case of corrins. Maximum turnover numbers obtained for corrins were about 100 product molecules/B12 molecule. Corrins have 6 double bonds and porphyrins contain 11 double bonds in their tetrapyrrole macrocycle. Consequently, corrins are less susceptible to proton attacks owing to their higher saturation as compared to that of porphyrins.

Armstrong and co-workers recently reported visible light-driven photocatalytic CO_2 reduction using a prototype enzyme-based system (Chaudhary et al. 2012; Woolerton et al. 2010). One of their pioneering work reports about the assembling of TiO_2 nanoparticles (NPs) and the CO_2-reducing enzyme Carbon monoxide dehydrogenase (CODH) I from the anaerobic microbe *Carboxydothermus hydrogenoformans (Ch).* Ch expresses five CODH complexes. CODH I, one among them, contains an unusual [Ni4Fe-4S] active site. The active site shows excellent reversibility in oxidation of CO to CO_2 (Wu et al. 2005) by just providing little overpotential (Parkin et al. 2007). In the presence of RuP sensitizer ([$Ru^{II}(bipy)_2(4,4'-(PO_3H_2)_2$-bipy)]$Br_2$; bipy=2,2′-bipyridine), this prototype enzyme-based system containing TiO_2 NPs and CODH I shows astonishing photocatalytic activity in reduction of CO_2 to CO (Figure 7.7a). Maximum CO yield obtained ca. 6 μmol in 4 h at 20°C, pH 6.0, and under exclusive visible light illumination. When the wide band gap semiconductor TiO_2 NPs were replaced

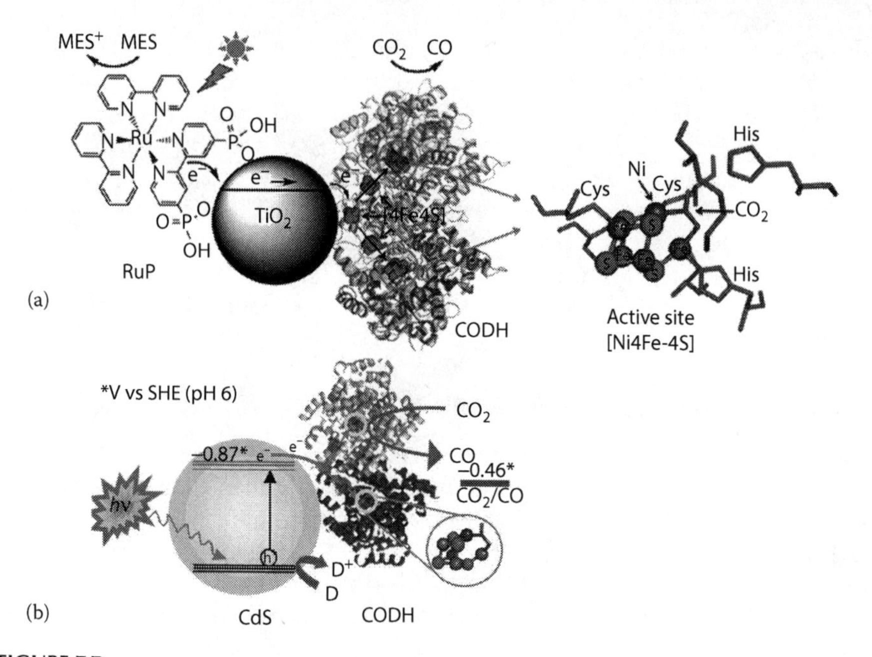

FIGURE 7.7
(a) Cartoon representation of the CO_2 photoreduction system using ch CODH I attached to RuP-modified TiO_2 NPs. A catalytic intermediate of the active site of the closely related enzyme CODH II with a bound substrate (CO_2, indicated with an arrow) (Data from Jeoung, J. H., and H. Dobbek, *Science*, 318, 1461–1464, 2007; Jeoung, J. H., and H. Dobbek, *Journal of American Chemical Society*, 131, 9922–9923, 2009) is also shown. The oxidized photosensitizer is recovered by the sacrificial electron donor MES. The enzyme structure used in the cartoon is CODH II, created using PyMOL. (Reprinted with permission from Woolerton, T. W. et al., *Journal of American Chemical Society*, 132, 2132–2133, 2010. Copyright [2010] American Chemical Society). (b) Schematic representation of visible light-driven CO_2 reduction using CODH-CdS nanocrystal assemblies. D represents an electron donor. (Chaudhary, Y. S. et al., *Chemical Communications*, 48, 58–60, 2012. Reproduced by permission of the Royal Society of Chemistry.)

with a narrower bandgap one, CdS nanocrystals, similar activity was obtained without using any sensitizer (Figure 7.7b) (Chaudhary et al. 2012).

Points to be noted here are that, however, the best selectivity can be obtained from the photocatalytic CO_2 reduction with these biomimetic/enzyme-based photocatalysts, and their stability in the ambient condition is the major concern and hinders this field to grow further.

7.4 Conclusion

Few simple, yet very crucial aspects have concisely been enlightened in this chapter. This will help enormously to understand the basic steps involved in the photocatalytic CO_2 reduction process and to design a simple, economical, and more efficient photocatalyst for CO_2 reduction. In addition, few more points are noteworthy to mention here. As compared to solar hydrogen generation, photocatalytic CO_2 reduction to solar fuels is more advantageous and, at the same time, far more challenging and less explored. The former occurs mainly in water, which eventually is the substrate itself to split into oxygen and hydrogen, the only two products in gaseous form. Therefore, separation of the products is not a concern. In contrast, photocatalytic CO_2 reduction can also be performed in aqueous medium, CO_2 continuously needs to be purged, and there are more chances of proton than CO_2 to be reduced as the latter's concentration is scarce owing to its poor solubility in water. This shortcoming can be avoided performing photocatalytic CO_2 reduction in the gas phase. As discussed in Section 7.2, photocatalytic CO_2 reduction often leads to a variety of products ranging from liquid to gaseous in phases. Therefore, one should be more cautious during the detection of the yield for the whole photocatalytic process. Sum must be taken for the amounts of all individual products after their careful identifications and quantifications. Also the control experiments without purging CO_2 or using ^{13}C labeled CO_2 need to be performed to exclude the possibilities of any misleading result coming out from any organic residual remaining on the photocatalyst itself.

References

Agarwal, J., E. Fujita, H. F. SchaeferIII, and J. T. Muckerman. 2012. Mechanisms for CO production from CO_2 using reduced rhenium tricarbonyl catalysts. *Journal of American Chemical Society* 134, no. 11: 5180–5186. doi: 10.1021/ja2105834.

Arai, T., S. Sato, K. Uemura, T. Morikawa, T. Kajino, and T. Motohiro. 2010. Photoelectrochemical reduction of CO_2 in water under visible-light

irradiation by a p-type InP photocathode modified with an electropolymerized ruthenium complex. *Chemical Communications* 46: 6944–6947. doi: 10.1039/c0cc02061c.
Bhugun, I., D. Lexa, and J. M. Saveant. 1994. Ultraefficient selective homogeneous catalysis of the electrochemical reduction of carbon dioxide by an iron(0) porphyrin associated with a weak bronsted acid cocatalyst. *Journal of American Chemical Society* 116, no. 11 (June): 5015–5017. doi: 10.1021/ja00090a068.
Bhugun, I., D. Lexa, and J. M. Saveant. 1996. Catalysis of the electrochemical reduction of carbon dioxide by iron(0) porphyrins: synergystic effect of weak bronsted acids. *Journal of American Chemical Society* 118, no. 7 (February): 1769–1776. doi: 10.1021/ja9534462.
Bonin, J., M. Chaussemier, M. Robert, and M. Routier. 2014. Homogeneous photocatalytic reduction of CO_2 to CO using iron(0) porphyrin catalysts: Mechanism and intrinsic limitations. *ChemCatChem* 6, no. 11 (November): 3200–3207. doi: 10.1002/cctc.201402515.
Bonin, J., M. Robert, and M. Routier. 2014. Selective and efficient photocatalytic CO_2 reduction to CO using visible light and an iron-based homogeneous catalyst. *Journal of the American Chemical Society* 136, no. 48 (November): 16768–16771. doi: 10.1021/ja510290t.
Bourrez, M., F. Molton, S. Chardon-Noblat, and A. Deronzier. 2011. [Mn(bipyridyl)$(CO)_3$Br]: an abundant metal carbonyl complex as efficient electrocatalyst for CO_2 reduction. *Angewandte Chemie International Edition English* 50, no. 42 (October): 9903–9906. doi: 10.1002/anie.201103616.
Chang, X., T. Wang, and J. Gongdoi. 2015. CO_2 photo-reduction: insights into CO_2 activation and reaction on surfaces of photocatalysts. *Energy & Environmental Science,* Advanced Article. doi: 10.1039/C6EE00383D.
Chaudhary, Y. S., T. W. Woolerton, C. S. Allen, et al. 2012. Visible light-driven CO_2 reduction by enzyme coupled CdS nanocrystals. *Chemical Communications* 48: 58–60. doi: 10.1039/c1cc16107e.
Cho, I. S., M. Logar, C. H. Lee, L. Cai, F. B. Prinz, and X. Zheng. 2014. Rapid and controllable flame reduction of TiO_2 nanowires for enhanced solar water-splitting. *Nano Letters* 14, no. 1: 24–31. doi: 10.1021/nl4026902.
Costentin, C., S. Drouet, M. Robert, and J.-M. Savéant. 2012. A local proton source enhances CO_2 electroreduction to CO by a molecular Fe catalyst. *Science* 338, no.6103 (October): 90–94. doi: 10.1126/science.1224581.
Dillip, G. R., A. N. Banerjee, V. C. Anitha, B. D. P. Raju, S. W. Joo, and B. Ki Min. 2017. Oxygen vacancy-induced structural, optical, and enhanced supercapacitive performance of zinc oxide anchored graphitic carbon nanofiber hybrid electrodes. *ACS Applied Materials & Interfaces* 8, no. 7 (February): 5025–5039. doi: 10.1021/acsami.5b12322.
Etacheri, V, M. K. Seery, S. J. Hinder, and S. C. Pillai. 2011. Oxygen rich titania: a dopant free, high temperature stable, and visible-light active anatase photocatalyst. *Advanced Functional Materials* 21 (September): 3744–3752. doi: 10.1002/adfm.201100301.
Fan, W., Q. Zhang, and Y. Wang. 2013. Semiconductor-based nanocomposites for photocatalytic H_2 production and CO_2 conversion. *Physical Chemistry Chemical Physics* 15: 2632–2649. doi: 10.1039/c2cp43524a.
Freund, H. J, and R. P. Messmer. 1986. On the bonding and reactivity of CO_2 on metal surfaces. *Surface Science* 172, no. 1: 1–30. doi: 10.1016/0039-6028(86)90580-7.

Freund, H. J., and M. W. Roberts. 1996. Surface chemistry of carbon dioxide. *Surface Science Reports* 25: 227–273. doi: 10.1016/S0167-5729(96)00007-6.

Fujiwara, H., H. Hosokawa, K. Murakoshi, et al. 1997. Effect of surface structures on photocatalytic CO_2 reduction using quantized CdS nanocrystallites. *Journal of Physical Chemistry B* 101, no. 41 (October): 8270–8278. doi: 10.1021/jp971621q.

Gattrell, M., N. Gupta, and A. Co. 2007. A review of the aqueous electrochemical reduction of CO_2 to hydrocarbons at copper. *Journal of Electroanalytical Chemistry* 594: 1–19. doi: 10.1016/j.jelechem.2007.05.013.

Gong, X. Q., and A. Selloni. 2005. Reactivity of anatase TiO_2 nanoparticles: the role of the minority (001) Surface. *Journal of Physical Chemistry B* 109, no. 42 (October): 19560–19562. doi: 10.1021/jp055311g.

Grodkowski, J., and P. Neta. 2000. Cobalt corrin catalyzed photoreduction of CO_2. *Journal of Physical Chemistry A* 104: 1848–1853. doi: 10.1021/jp9939569.

Habisreutinger, S. N., L. Schmidt-Mende, and J. K. Stolarczyk. 2013. Photocatalytic reduction of CO_2 on TiO_2 and other semiconductors. *Angewandte Chemie International Edition* 52: 7372–7408. doi: 10.1002/anie.201207199.

Han, X., Q. Kuang, M. Jin, Z. Xie, and L. Zheng. 2009. Synthesis of titania nanosheets with a high percentage of exposed (001) facets and related photocatalytic properties. *Journal of American Chemical Society* 131, no. 9: 3152–3153. doi: 10.1021/ja8092373.

Hawecker, J., Lehn, J.-M., and R. J. Ziessel. 1983. Efficient photochemical reduction of CO_2 to CO by visible light irradiation of systems containing Re(bipy)$(CO)_3X$ or $Ru(bipy)_3^{2+}$–Co^{2+} combinations as homogeneous catalysts. *Journal of the Chemical Society, Chemical Communication* no. 9: 536–538. doi: 10.1039/C39830000536.

Indrakanti, V. P., J. D. Kubickib, and H. H. Schobert. 2009. Photoinduced activation of CO_2 on Ti-based heterogeneous catalysts: Current state, chemical physics-based insights and outlook. *Energy & Environmental Science* 2: 745–758. doi: 10.1039/b822176f.

Inoue, T., A. Fujishima, S. Konishi, and K. Honda. 1979. Photoelectrocatalytic reduction of carbon dioxide in aqueous suspensions of semiconductor powders. *Nature* 277: 637–638. doi: 10.1038/277637a0.

Jeletic, M. S., M. T. Mock, A. M. Appel, and J. C. Linehan. 2013. A cobalt-based catalyst for the hydrogenation of CO_2 under ambient conditions. *Journal of the American Chemical Society* 135, no. 31 (August): 11533–11537. doi: 10.1021/ja406601v.

Jeoung, J. H., and H. Dobbek. 2007. Carbon dioxide activation at the Ni, Fe-cluster of anaerobic carbon monoxide dehydrogenase. *Science* 318, no. 5855: 1461–1464. doi: 10.1126/science.1148481.

Jeoung, J. H., and H. Dobbek. 2009. Structural basis of cyanide inhibition of Ni, Fe-containing carbon monoxide dehydrogenase. *Journal of American Chemical Society* 131, no. 29: 9922–9923. doi: 10.1021/ja9046476.

Kanemoto, M., H. Hosokawa, Y. Wada, et al. 1997. Semiconductor photocatalysis. Part 20. Role of surface in the photoreduction of carbon dioxide catalysed by colloidal ZnS nanocrystallites in organic solvent. *Journal of the Chemical Society, Faraday Transactions* 92: 2401–2411. doi: 10.1039/FT9969202401.

Kong, M., Y. Li, X. Chen, et al. 2011. Tuning the relative concentration ratio of bulk defects to surface defects in TiO_2 nanocrystals leads to high photocatalytic efficiency. *Journal of American Chemical Society* 133: 16414–16417. doi: 10.1021/ja207826q.

Kuriki, R., K. Sekizawa, O. Ishitani, and K. Maeda. 2015. Visible-light-driven CO_2 reduction with carbon nitride: enhancing the activity of ruthenium catalysts. *Angewandte Chemie International Edition* 54, no. 8 (February): 2406–2409. doi: 10.1002/anie.201411170.

Lazzeri, M., A. Vittadini, and A. Selloni. 2001. Charge pumping and photovoltaic effect in open quantum dots. *Physical Review B* 63: 55409. doi: http://dx.doi.org/10.1103/PhysRevB.63.195313.

Lazzeri, M., A. Vittadini, and A. Selloni. 2002. Structure and energetics of stoichiometric TiO_2 anatase surfaces. *Physical Review B* 65: 119901. doi: http://dx.doi.org/10.1103/PhysRevB.63.155409.

Lee, J., D. C. Sorescu, and X. Deng. 2011. Electron-induced dissociation of CO_2 on TiO_2 (110). *Journal of American Chemical Society* 133, 10066–10069. doi: 10.1021/ja204077e.

Lehn, J.-M., and R. Ziessel. 1982. Photochemical generation of carbon monoxide and hydrogen by reduction of carbon dioxide and water under visible light irradiation. *Proceedings of the National Academy of Sciences USA* 79 (January): 701–704. PMCID: PMC345815.

Li, M., L. Zhang, X. Fan, Y. Zhou, M. Wu, and J. Shi. 2015. Highly selective CO_2 photoreduction to CO over g-C_3N_4/Bi_2WO_6 composites under visible light. *Journal of Materials Chemistry A* 3: 5189–5197. doi: 10.1039/c4ta06295g.

Liao, W., M. Murugananthan, and Y. Zhang. 2015. Synthesis of Z-scheme g-C_3N_4–Ti^{3+}/TiO_2 material: an efficient visible light photoelectrocatalyst for degradation of phenol. *Physical Chemistry Chemical Physics* 17, no. 14: 8877–8884. doi: 10.1039/C5CP00639B.

Liao, Y., S.-W. Cao, Y. Yuan, Q. Gu, Z. Zhang, and C. Xue. 2014. Efficient CO_2 capture and photoreduction by amine-functionalized TiO_2. *Chemistry* 20, no. 33 (August): 10220–10222. doi: dx.doi.org/10.1002/chem.201403321.

Lin, J., Z. Pan, and X. Wang. 2014. Photochemical reduction of CO_2 by graphitic carbon nitride polymers. *ACS Sustainable Chemistry & Engineering* 2, no. 3 (December): 353–358. doi: 10.1021/sc4004295.

Liu, L, H. Zhao, J. M. Andino, and Y. Li. 2012. Photocatalytic CO_2 reduction with H_2O on TiO_2 nanocrystals: comparison of anatase, rutile, and brookite polymorphs and exploration of surface chemistry. *ACS Catalysis* 2 (July): 1817–1828. doi: 10.1021/cs300273q.

Meng, X., S. Ouyang, T. Kako, et al. 2014. Photocatalytic CO_2 conversion over alkali modified TiO_2 without loading noble metal cocatalyst. *Chemical Communications* 50: 11517–11519. doi: 10.1039/c4cc04848b.

Morris, A. J., G. J. Meyer, and E. Fujita. 2009. Molecular approaches to the photocatalytic reduction of carbon dioxide for solar fuels. *Accounts of Chemical Research* 42, no. 12 (December): 1983–1994. doi: 10.1021/ar9001679.

Munoz-Batista, M. J., A. Kubacka, and M. Fernández-García. 2014. Effect of g-C_3N_4 loading on TiO_2-based photocatalysts: UV and visible degradation of toluene. *Catalysis Science & Technology* 4, no. 7: 2006–2015. doi: 10.1039/C4CY00226A.

Parkin, A., J. Seravalli, K. A. Vincent, S. W. Ragsdale, and F. A. Armstrong. 2007. Rapid and efficient electrocatalytic CO_2/CO interconversions by *carboxydothermus hydrogenoformans* CO dehydrogenase I on an electrode. *Journal of American Chemical Society* 129, no. 34: 10328–10329. doi: 10.1021/ja073643o.

Qu, Y. Q., and X. F. Duan. 2013. Progress, challenge and perspective of heterogeneous photocatalysts. *Chemical Society Reviews* 42: 2568–2580. doi: 10.1039/c2cs35355e.

Reithmeier, R., C. Bruckmeier, and B. Rieger. 2012. Conversion of CO_2 via visible light promoted homogeneous redox catalysis. *Catalysts* 2, no. 4 (November): 544–571. doi: 10.3390/catal2040544.

Sato, S., T. Arai, T. Morikawa, et al. 2011. Selective CO_2 conversion to formate conjugated with H_2O oxidation utilizing semiconductor/complex hybrid photocatalysts. *Journal of American Chemical Society* 133, no. 39: 15240–15243. doi: 10.1021/ja204881d.

Sato, S., T. Morikawa, S. Saeki, T. Kajino, and T. Motohiro. 2010. Visible-light-induced selective CO_2 reduction utilizing a ruthenium complex electrocatalyst linked to a p-type nitrogen-doped Ta_2O_5 Semiconductor. *Angewandte Chemie International Edition* 49: no. 30 (July): 5101 –5105. doi: 10.1002/anie.201000613.

Schmidt, M. 1994. The thermodynamics of CO_2 conversion. In *Carbon dioxide Chemistry: Environmental Issues*, eds. J. P. Paul, and Claire-Marie, 22–30. Cambridge: The Royal Society of Chemistry.

Sekizawa, K., K. Maeda, K. Domen, K. Koike, and O. Ishitani. 2012. Artificial Z-scheme constructed with a supramolecular metal complex and semiconductor for the photocatalytic reduction of CO_2. *Journal of American Chemical Society* 135, no. 12: 4596–4599. 10.1021/ja311541a.

Tamaki, Y., T. Morimoto, K. Koike, and O. Ishitani. 2012. Photocatalytic CO_2 reduction with high turnover frequency and selectivity of formic acid formation using Ru(II) multinuclear complexes. *Proceedings of the National Academy of Sciences USA* 109, no. 39: 15673–15678. doi: 10.1073/pnas.1118336109.

Tan, L.-L., W.-J. Ong, S.-P. Chai, and A. R. Mohamed. 2014. Band gap engineered, oxygen-rich TiO_2 for visible light induced photocatalytic reduction of CO_2. *Chemical Communications* 50, 6923–6927. doi: 10.1039/c4cc01304b.

Thoi, V. S., N. Kornienko, C. G. Margarit, P. Yang, and C. J. Chang. 2013. Visible-light photoredox catalysis: selective reduction of carbon dioxide to carbon monoxide by a nickel N-heterocyclic carbene-isoquinoline complex. *Journal of the American Chemical Society* 135, no.38 (September): 14413–14424. doi: 10.1021/ja4074003.

Tu, W., Y. Zhou, and Z. Zou. 2014. Photocatalytic conversion of CO_2 into renewable hydrocarbon fuels: state-of-the-art accomplishment, challenges, and prospects. *Advanced Materials* 26: 4607–4626. doi: 10.1002/adma.201400087.

Wang, J., H.-C. Yao, Z.-Y. Fan, et al. 2016. Indirect Z-scheme BiOI/g-C_3N_4 photocatalysts with enhanced photoreduction CO_2 activity under visible light irradiation. *ACS Applied Materials & Interfaces* 8, no. 6: 3765–3775. doi: 10.1021/acsami.5b09901.

Wang, J.-C., L. Zhang, W.-X. Fang, et al. 2015. Enhanced photoreduction CO_2 activity over direct Z-Scheme α-Fe_2O_3/Cu_2O heterostructures under visible light irradiation. *ACS Applied Materials & Interfaces* 7, no. 16: 8631–8639. doi: 10.1021/acsami.5b00822.

Wen, C. Z., J. Z. Zhou, H. B. Jiang, Q. H. Hu, S. Z. Qiao, and H. G. Yang. 2011. Synthesis of micro-sized titanium dioxide nanosheets wholly exposed with high-energy {001} and {100} facets. *Chemical Communications* 47: 4400–4402. doi: 10.1039/C0CC05798C.

White, T. A., S. Maji, and S. Ott. 2014. Mechanistic insights into electrocatalytic CO_2 reduction within [RuII(tpy)(NN)X]$^{n+}$ architectures. *Dalton Transactions* 43, no. 40: 15028–15037. doi: 10.1039/c4dt01591f.

Woolerton, T. W., S. Sheard, E. Reisner, E. Pierce, S. W. Ragsdale, and F. A. Armstrong. 2010. Efficient and Clean photoreduction of CO_2 to CO by enzyme-modified TiO_2. *Journal of American Chemical Society* 132, no. 7: 2132–2133. doi: 10.1021/ja910091z.

Wu, M., Q. Ren, A. S. Durkin, et al. 2005. Life in hot carbon monoxide: The complete genome sequence of carboxydothermus hydrogenoformans Z-2901. *PloS Genetics* 1: 563–574. http://dx.doi.org/10.1371/journal.pgen.0010065.

Xie, S., Y. Wang, Q. Zhang, W. Fan, W. Deng, and Y. Wang. 2013. Photocatalytic reduction of CO_2 with H_2O: significant enhancement of the activity of Pt–TiO_2 in CH_4 formation by addition of MgO. *Chemical Communications* 49: 2451–2453. doi: 10.1039/c3cc00107e.

Yang, J., D. Wang, H. Han, and C. Li. 2013. Roles of cocatalysts in photocatalysis and photoelectrocatalysis. *Accounts of Chemical Research* 46, no. 8 (March): 1900–1909. doi: 10.1021/ar300227e.

Yu, J., J. Low, W. Xiao, P. Zhou, and M. Jaroniec. 2014. Enhanced photocatalytic CO_2-reduction activity of anatase TiO_2 by co-exposed {001} and {101} facets. *Journal of American Chemical Society* 136, no. 25: 8839–8842. doi: 10.1021/ja5044787.

Yu, J., S. Wang, J. Low, and W. Xiao. 2013. Enhanced photocatalytic performance of direct Z-scheme g-C_3N_4–TiO_2 photocatalysts for the decomposition of formaldehyde in air. *Physical Chemistry Chemical Physics* 15: 16883–16890. doi: 10.1039/C3CP53131G.

Zhou, P., J. Yu, and M. Jaroniec. 2014. All-solid-state Z-scheme photocatalytic systems. *Advanced Materials* 26: 4920–4935. doi: 10.1002/adma.201400288.

Zhou, S., Y. Liu, J. Li, et al. 2014. Facile *in situ* synthesis of graphitic carbon nitride (g-C_3N_4)-N-TiO_2 heterojunction as an efficient photocatalyst for the selective photoreduction of CO_2 to CO. *Applied Catalysis B: Environmental* 158–159: 20–29. doi: 10.1016/j.apcatb.2014.03.037.

Index

D

Q

R

S